The Institution of StructuralEngineers

The future of structural design

This version (1.0) published March 2025 by
The Institution of Structural Engineers
International HQ
47–58 Bastwick Street
London, EC1V 3PS
United Kingdom
www.istructe.org

Typeset by Riverside Publishing Solutions Ltd

Printed and bound by Hobbs The Printers Ltd

WORLD LAND TRUST™
www.carbonbalancedprint.com
CBP2250

MIX
Paper | Supporting responsible forestry
FSC® C020438
www.fsc.org

ISBN (Print) 978-1-906335-78-6
ISBN (PDF) 978-1-906335-79-3
ISBN (ePUB) 978-1-906335-81-6

Contents

PART FIVE

The future of structural design

PART SIX

Appendices

Acknowledgements

The Institution would like to thank Jenny Spratt (J. Spratt Design) for creating the visual approach for this book.

The authors would like to thank the following reviewers for their helpful feedback across multiple drafts of this work:

Will Arnold (The Institution of Structural Engineers)
Steve Denton (WSP)
Joseph Donohue (Arup)
Nick Eckford
Gareth Evans (Constructex)
Donya Hajializadeh (University of Surrey)
Patrick Hayes (The Institution of Structural Engineers)
Peter Ho (Dragon Engineers)
Robert Jones (AKT II)

Permission to reproduce the following has been obtained, courtesy of these individuals/organisations:

Full page imagery
Page xiv: ©Ed Hawkins. Licensed under Creative Commons Attribution 4.0 International
Pages 38, 110, 222 and 282: ©James McKay and The Institution of Structural Engineers. Licensed under Creative Commons Attribution 4.0 International
Page 332: ©James McKay

Figures
Figure 1.2: ©Jan Konietzko. Licensed under Creative Commons Attribution 4.0 International
Figure 1.3: Adapted from ©Jan Konietzko. Licensed under Creative Commons Attribution 4.0 International
Figure 4.1: ©Milkwood

Figure 4.3: ©European Union. Licensed under Creative Commons Attribution 4.0 International
Figure 5.1: ©Think Up. Licensed under Creative Commons Attribution-ShareAlike 4.0 International
Figure 5.4: ©Neil Tutt
Figures 5.5, 11.11, 11.24 and 14.4: ©Integral Engineering Design
Figure 7.1: ©Steve Denton/WSP
Figure 10.17: Derived/adapted from RICS *Whole life carbon assessment (WLCA) for the built environment (2nd edition)*, BS EN 15978, BS EN 17472 and BS EN 15643
Figures 11.1 and 11.19 (middle): ©Ramboll
Figure 11.2: ©Simon Smith/Smith and Wallwork
Figure 11.3: ©Sandberg
Figure 11.4: ©Dominika Malkowska. Licensed under Creative Commons Attribution 4.0 International Deed
Figure 11.6: ©Natasha Watson/Buro Happold
Figure 11.12: ©Heyne Tillett Steel
Figures 11.14 and 11.17 (right): ©Ellen Grist/Ramboll/University of Bath
Figures 11.15 and 11.16: ©Alex Lynes/Webb Yates Engineers
Figure 11.18: ©The Institution of Structural Engineers and Warringtonfire Testing and Certification Ltd
Figure 11.19 (left): ©Joshua Mudie
Figure 11.20: Permission obtained from Frei Otto Archives (SAAI)
Figure 11.23: ©The UKCRIC Soil-Foundation-Structure-Interaction (SoFSi) Laboratory
Figure 13.1(a): ©Bristol Region Building Record, Archives & Research Collections, University of Bath
Figure 13.1(b): Wikipedia Commons ©Stephen Richards (CC BY-SA 2.0)
Figure 13.2: Perry, J.G. *Report R 133: A Guide to the Management of Building Refurbishment*
©CIRIA. Reproduced with permission of the Licensor through PLSclear

Figure 14.1: Reproduced with the permission of the National Library of Scotland
Figure 14.2: ©The John Gent Collection, CNHSS
Figure 14.5: ©Above Horizon Photography
Figures 17.1–17.4: Reprinted from 'Eurocode-based seismic assessment of modern heritage RC structures: The case of the Tower of the Nations in Naples (Italy)' by De Luca, F., Verderame, G.M. and Manfredi, G. in *Engineering Structures*, Vol.74, pp.96–110, 2014.
With permission from Elsevier
Figure C.1: ©Veronika Turcanova/University of Bristol

Tables

Tables 6.4 and 6.5: With permission from ASCE
Table 6.6: Used with permission of Elsevier BV, from 'Developments towards full probabilistic design codes' by Vrouwenvelder, A.C.W.M., in *Structural safety* 24(2-4), pp.417-432, 1982. Permission conveyed through Copyright Clearance Center, Inc.
Table 13.2: ©European Union. From Luechinger P. et al. *New European Technical Rules for the Assessment and Retrofitting of Existing Structures.* January 2015.
Table C.1: ©James Norman and Joel Ross (University of Bristol)

Figures 10.17, 13.3, 13.4 and Tables 5.1, 5.2, 6.1-6.3, 13.3 and 17.1: Permission to reproduce extracts from British Standards is granted by BSI. British Standards can be obtained in PDF or hard copy formats from BSI Knowledge: https://knowledge.bsigroup.com or by contacting BSI Customer Services for hardcopies only: Tel: +44 (0)20 8996 9001, Email: cservices@bsigroup.com.

The authors

JAMES NORMAN

James has been a structural engineer all his life. He studied Civil Engineering at the University of Nottingham, joined Whitbybird in 2000, left to do a PhD at the University of Bristol in structural dynamics between 2003–6, and rejoined Whitbybird (who became Ramboll soon after) in 2006. James however, continued to teach structural design one day per week. In 2011 he started teaching a unit on sustainable construction and in 2012 joined Integral Engineering Design to get more hands-on experience in sustainable design (including rammed earth, straw-bale, timber and reuse).

James has designed using a number of different materials including steel, concrete, timber, masonry, earth, straw and glass. He has worked on a wide variety of projects; from the façade of the Tate Modern extension to a small glass table at the Tate St Ives.

In 2015 James became a full time academic but continues to think and behave like a structural engineer. James supervises a number of research projects from Undergraduate to PhD and most include physical testing of one form or another. He has published experimental work on bamboo, timber concrete composites and lime concrete. He teaches design to all years for half the year and during the other half he writes books and articles for structural engineers including The Regenerative Structural Engineer (IStructE) and Conceptual Design of Buildings (IStructE). In 2020 James was awarded a National Teaching Fellowship for his innovative work in teaching and became Professor of Sustainable Design. In 2024 he was elected as a Fellow of the IStructE through the Eminent Person's route.

DAN BERGSAGEL

Dan Bergsagel is the Sustainability Lead in schlaich bergermann partner's New York City office. He is also a Visiting Scholar with the Circular Construction Lab in the Cornell University College of Architecture, Art and Planning; an Adjunct Professor in structures at Kean University's School of Public Architecture; and a co-founder of educational non-profit Scale Rule. Dan previously designed pedestrian bridges, stadia and art installations from NYC, and buildings and towers from London.

He studied at the University of Cambridge and École Centrale Paris, and has previously worked and taught in London at AKT II, University College London and the Architectural Association. He is a member of the SEI Sustainability Committee and Chair of the committee's Circular Economy Working Group, a member of the IStructE Sustainability Panel, and co-founded RECLAIM NYC, an interdisciplinary group collaborating to develop circular economy policy proposals for New York City's construction and building removal industry.

SCOTT BOOTE

As a Chartered Structural Engineer with 20+ years of experience, I'm dedicated to shaping a more sustainable future in construction. My passion lies in architecturally led structures, collaborating with design teams to create innovative, low-carbon solutions that are both beautiful and functional.

Specialising in low-carbon materials, I've pioneered the use of structural stone, rammed earth, and timber-concrete composites. With 60+ completed structural stone projects, including post-tensioned systems, I'm one of the few engineers globally with this level of expertise, earning an

IStructE Award in 2016. I then went on to author a well-received series of articles for *The Structural Engineer* magazine on the use of stone as a structural material.

Whether designing new-builds, sculptural staircases, or complex refurbishments, I push boundaries to create lasting, sustainable structures that challenge conventional materials and methods.

ALIX DIETZEL

Dr Alix Dietzel is a Senior Lecture in Climate Justice at SPAIS, University of Bristol. She has 14 years of academic experience and has focused mainly on ethical aspects of climate policy and more recently just transition, focusing both on the global and local scale. She also sits on the Bristol Advisory Committee on Climate Change and is the Associate Director for Impact and Innovation at the Cabot Institute for the Environment. Alix works across disciplines, and her main focus has been on the importance of inclusive decision making and ensuring that the most affected are heard.

FLAVIA DE LUCA

Flavia De Luca is Professor of Structural and Earthquake Engineering at the University of Bristol. She received her PhD in Seismic Risk at the University of Naples Federico II in 2012. She worked for two years as post-doctoral researcher at the University of Naples Federico II and joined the University of Bristol as Lecturer in 2014. De Luca has been working over the last fifteen years in earthquake engineering and disaster risk reduction. Over the last five years, she has been part of international programmes related to risk and resilience in different countries. She has led projects with strong experimental components and advanced analytical and numerical work in the field of structural earthquake engineering. Her expertise spans nonlinear modelling of concrete structures, soil-structure interaction, large-scale experimental testing, risk assessment and structural vulnerability.

PHIL ISAAC

Phil is co-founder of the socially-focused structural engineering consultancy Simple Works, and community interest company Scale Rule CIC. He is a Visting Lecturer at the University of Bath, where he obtained his PhD in Civil Engineering, and is a Fellow of the Institution of Structural Engineers. He co-founded Simple Works in 2019 to focus on projects with society and the environment at their heart. He leads the research agenda in the practice, which includes a focus on regenerative design principles — particularly low carbon materials and circular construction.

GAVIN KNOWLES

Gavin is a lecturer at the University of Bath. Prior to this he worked in practice as a Chartered structural engineer. His previous projects include many education and office buildings, along with conservation and refurbishment projects — interweaved with diverse structures such as rammed chalk-walled houses, recycled material stages at WOMAD Festival and the odd sculpture. He now teaches Structural Design, Conservation and leads on student design projects.

DAN MASKELL

Dr. Dan Maskell is a Senior Lecturer in the Department of Architecture and Civil Engineering at the University of Bath, specializing in innovative natural building materials and their integration into mainstream construction. His research takes a holistic approach, examining these materials' structural performance, indoor environmental regulation, and impact on occupant health and well-being. His work ranges from modifying materials at the atomic level to monitoring full-scale buildings. Through this comprehensive approach, he aims to advance sustainable building practices, ensuring the built environment not only minimizes its impact but also contributes to the regeneration of the natural ecosystem.

RAFFAELE DE RISI

Dr De Risi is an associate professor of civil engineering at the School of Civil, Aerospace, and Design Engineering, University of Bristol. He received his BSc degree in Civil Engineering and MSc in Structural and Geotechnical Engineering from the University of Naples Federico II (Italy) in 2007 and 2009, respectively. In 2013, he was awarded a PhD in Materials Engineering and Structures at the University of Naples Federico II (Italy). His research interests are broad and multidisciplinary and cover a wide range of academic fields, including structural reliability, engineering seismology, earthquake engineering, tsunami engineering, flood risk, and decision-making under uncertainty. His research spans several countries, including Canada, Ethiopia, Indonesia, Italy, Japan, Malawi, Nepal, Tanzania, and the United Kingdom.

Part 1
Introduction

The world is changing. Global temperature is rising[P1.1], and biodiversity is declining[P1.2].

The power of computation is rapidly increasing, with new developments occurring more rapidly[P1.3].

What does all this mean for structural engineers?

Will we even need structural engineers in the future, or will AI replace us[P1.4]?

In this book, we look at the future of design. We examine how the role of engineers is already changing as we design for a low- and zero-carbon world, a circular economy and a regenerative future. We will explore how to understand uncertainty and reliability to tackle these problems more effectively.

Finally, we look further into the future. As the speed and power of computing increase, the ability to carry out complex analysis will move from the realm of academics and specialists and become within reach of all engineers. This will open up more complicated and nuanced design fields based on mathematical risk models.

What is the future of structural design, and what will the role of the structural engineer be in this future?

REFERENCES

P1.1 Hawkins, E. *Show Your Stripes*, National Centre for Atmospheric Science, University of Reading, 2023. Available at: https://showyourstripes.info/ [Accessed: January 2025].

P1.2 Kolbert, E. *The Sixth Extinction: An Unnatural History*. London: Bloomsbury Publishing, London, 2014.

P1.3 Moore, G. *Moore's Law*. Available at: https://en.wikipedia.org/wiki/Moore%27s_law [Accessed: January 2025].

P1.4 The Institution of Structural Engineers. *Computational engineering*. London: IStructE Ltd, 2020.

CHAPTER 1

Introduction

James Norman

1.1 EMERGENCY RESPONSE

In May 2019, the Architects Declare movement was born. This was swiftly followed by Structural Engineers Declare and other engineering declarations. The Institution of Structural Engineers (IStructE) has been at the forefront in the response to these industry declarations. They started by focusing on the climate emergency and how we can reduce embodied carbon in new construction projects. They produced articles, talks and blogs on reducing carbon, which led to the publication of *How to calculate embodied carbon*[1.1] and *Design for zero*[1.2].

Around the same time, the IStructE started to champion the area of reuse, which led to further conversations and articles[1.3, 1.4], and the publication of *Circular economy and reuse: guidance for designers*[1.5]. Simultaneously, people started exploring other approaches to sustainability and design. How can we go beyond 'do no harm' and create ways for humans and living systems to survive, thrive and co-evolve? This approach challenges us to rethink engineering practice, re-imagine our systems and supply chains, and think not about individual projects but how our whole industry needs to change — an approach generally referred to as 'regenerative design'. The IStructE published a whole issue of *The Structural Engineer* dedicated to the topic of regenerative design[1.6] (Figure 1.1) and went on to publish *The regenerative structural engineer*[1.7].

Figure 1.1: Cover of *The Structural Engineer*, August 2023 'Designing for life'

As engineers, we now need to be able to design considering a multiplicity of different framings while also ensuring our design is safe. The relationship between these different framings — sustainable, circular and regenerative, has been summarised in Figure 1.2[1.8].

Figure 1.2: Relationship between sustainability, circular and regenerative approaches

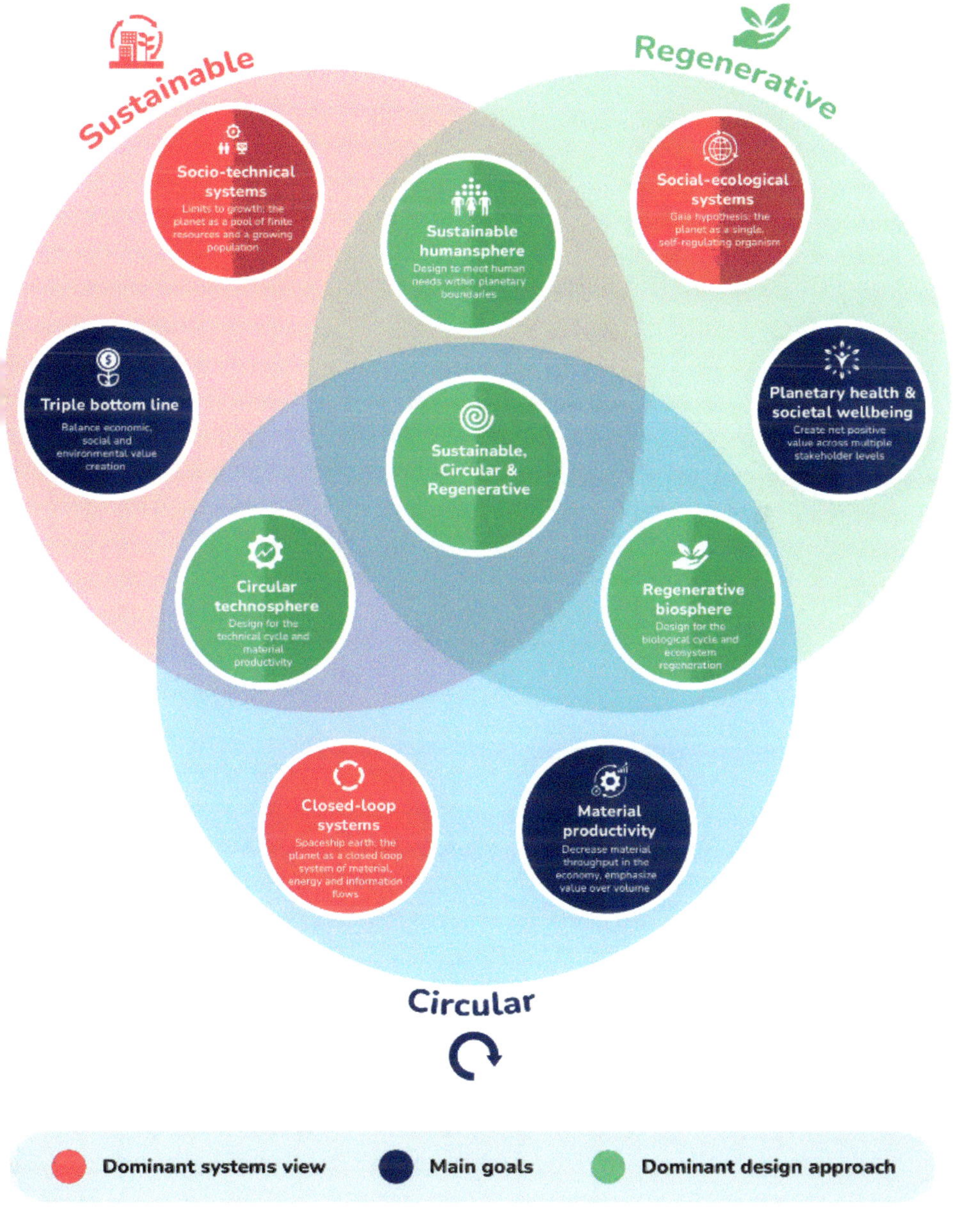

Each of these approaches requires us to have a profound understanding of design. Not just the mechanics but also uncertainty and how it is managed safely and efficiently. We need to understand uncertainty — when making designs as efficient as possible, when exploring reuse or when exploring new or local materials and characterising and testing them. This book aims to unpack the realities of design by studying uncertainties, underlying risk, and related subjects which also need to be considered.

1.2 STRUCTURE OF THE BOOK

> *"Uncertainty demands caution, and caution generates waste. So what are our sources of uncertainty, and how do we understand and control them?"* Steve Denton, Chair of CEN/TC 250

All three design approaches, sustainable, circular and regenerative carry uncertainty. Throughout Parts 2–4, we will explore these uncertainties as comprehensively as possible (Figure 1.3).

Sustainable design requires us to minimise waste and avoid over-design. In Part 2, we will look in detail at how uncertainties are considered in modern codes such as the Eurocodes and why we can minimise waste if we can control our uncertainty. Part 2 also serves as an introduction to reliability, discussing why we should design to current codes…

> "…*and no more."* Chris Wise, Expedition Engineering[1.9]

Regenerative design calls for use of both novel and local materials (while noting that not all novel and local materials are regenerative). It requires us to consider the different aspects of materials and use them optimally. In Part 3, we look at classifying materials and physically testing them.

We explore various models for expressing uncertainty (distributions), how to apply them and how confident we can be in them.

Circular design calls for us to reuse what we already have. In Part 4, we look at how we can begin to assess existing buildings and what information we already know about the building (after all, much of the uncertainty we have when designing a building is around how it will be built; once built, the nature of the uncertainty changes). We explore how we can update our understanding of uncertainty as we gather information and how, through careful probabilistic design, we can assess the adequacy of the existing materials, components and structural systems.

In Part 5, we turn our attention to the **future of design**. Where is all this leading? We explore risk-based design and how we can use probabilistic approaches to understand and balance consequences and available resources better.

ure 1.3: The structure of the book

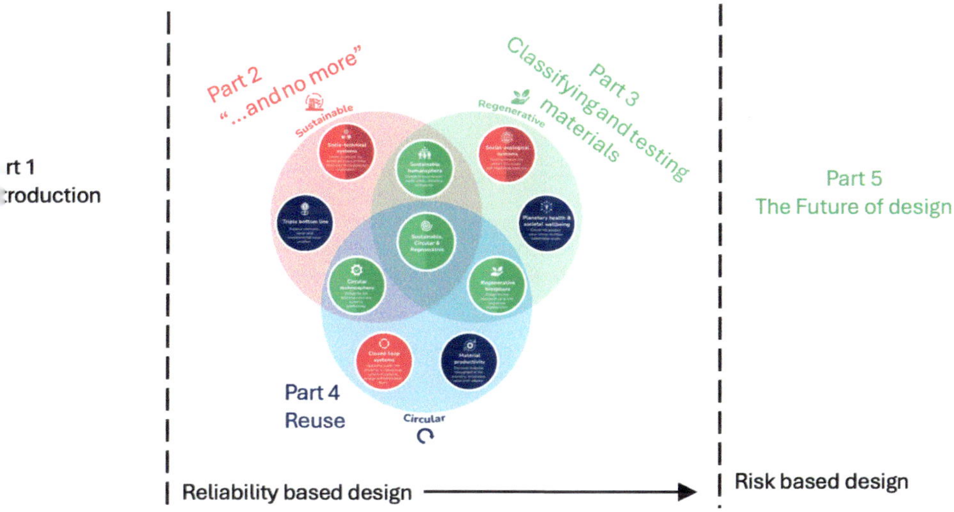

The remainder of Part 1 focuses on three topics. First, we provide a primer on risk- and reliability-based design. Second, we consider the ethics of design and why we should aim to minimise waste; in other words, why we should care about all of this. Third, we consider the history of our codes and reliability — where we have been, before we consider where we are going as we explore the future of design.

1.3 THOUGHT LEADERSHIP

At the start of the book, we want to be clear: this book is not a guide or a 'how-to'. While the ideas we present are already utilised in parts of our discipline, such as earthquake engineering or offshore design, they are not yet generally commonplace across structural engineering. Furthermore, each of these topics could be a standalone book. Therefore, this book is aimed to be more of a 'what if' than a guide. Where possible, we will provide references and further reading to help readers who want to delve deeper into the subjects. We have also provided some case studies, which we hope readers will replicate and play with. These should not be seen as design tools but more as preliminary explorations to see the potential these approaches could unlock. They should not be applied directly to real projects without rigorous consideration and review.

1.4 WARNING

This book looks to challenge engineers to engage more fully in understanding and using probabilistic methods in design. However, it does not suggest that engineers can define their own acceptable risk of failure; this is determined at a societal level and is defined in our design codes. Instead, this book challenges us to understand uncertainty better, including the society-defined acceptable risks of failure, and therefore, reduce waste by not unnecessarily going beyond these limits.

REFERENCES

1.1 The Institution of Structural Engineers. *How to calculate embodied carbon (Third edition).* London: IStructE Ltd, 2025.

1.2 The Institution of Structural Engineers. *Design for zero.* London: IStructE Ltd, 2021.

1.3 The Institution of Structural Engineers. 'How can we create an engineering industry while building nothing?' *The Structural Engineer*, 98(7), July 2020, p31. Available at: https://doi.org/10.56330/PAYC4580 [Accessed: January 2025].

1.4 The Institution of Structural Engineers. 'Nothing is better than something.' *The Structural Engineer*, 98(6), June 2020, p12. Available at: https://doi.org/10.56330/NYMP4183 [Accessed: January 2025].

1.5 The Institution of Structural Engineers. *Circular economy and reuse: guidance for designers.* London: IStructE Ltd, 2023.

1.6 The Institution of Structural Engineers. 'Complete issue: Designing for life'. *The Structural Engineer*, 101(8), August 2023. Available at: https://www.istructe.org/journal/volumes/volume-101-(2023)/issue-8/ [Accessed: January 2025].

1.7 The Institution of Structural Engineers. *The regenerative structural engineer.* London: IStructE Ltd, 2024.

1.8 Konietzko, J *et al.* 'Towards Regenerative Business Models: A Necessary Shift?' *Sustainable Production and Consumption*, 38, June 2023, pp372–388. Available at: https://doi.org/10.1016/j.spc.2023.04.014 [Accessed: January 2025].

1.9 Wise, C. 'What If Everything We Did Was Wrong?' *Building*, June 2010. Available at: https://www.building.co.uk/comment/what-if-everything-we-did-was-wrong/5000493.article [Accessed: January 2025].

CHAPTER 2

Introduction to risk and reliability

James Norman and Raffaele De Risi

Eurocode — Basis of structural design[2.1] defines reliability as the:

> *"ability of a structure or a structural member to fulfil the specified requirements, including the design working life, for which it has been designed. Reliability is usually expressed in probabilistic terms. NOTE Reliability covers safety, serviceability and durability of a structure."*

Carrying out a codified design is achieved through the application of partial safety factors using what is known as the semi-probabilistic design approach, also known as limit state design (LSD) or load and resistance factor design (LRFD).

We also consider risk-based design. In risk-based design, we equate consequences (failure) with opportunities.

Design has often focused on optimising several competing requirements. These requirements are to make our buildings safe, functional and affordable, and these are summarised according to the following (simplified) relationship:

$$min\ C_{Tot} = C_{Con} + P_F C_F \qquad 2.1$$

In risk-based design, we want to achieve the minimum (*min*) total cost (C_{Tot}) along the asset life cycle. To do this, we must consider the cost of construction (C_{Con}), the probability of failure (P_F) and the cost of failure (C_F). This is at the base of the life cycle costs (LCC) assessment, weighing up risk (i.e., the building failing) with opportunity (i.e., the ability to build cheaply). The previous formulation can be enriched by adding inspection and maintenance costs as well as other components that are beyond the scope of this book.

In our current paradigm of continuous economic growth, ultimately this cost will be counted in financial terms. As we transition to a new paradigm of holism, we may choose other ways to measure cost such as human and living system well-being[2.2].

Reliability looks purely at the probability of failure. It draws a line and says anything on one side of this line is safe; anything on the other is not safe. The definition of safe, far from simple, is a complex mixture of what is socially, ethically and economically viable. For reliability, we can write a similar (simplified) mathematical statement to that for risk:

$$min\ C_{Tot} = C_{Con} \mid P_F < P_{Ft} \qquad 2.2$$

In other words, we want to achieve minimum cost but that is conditional (sign |) on ensuring that the probability of failure (P_F) is below an acceptable limit (P_{Ft}).

Designs accounting for uncertainties are categorised into four levels:

Level I: uncertainties are modelled by a characteristic/nominal value, and structural safety is guaranteed via partial safety factors provided by building codes. This approach is also known as semi-probabilistic and is the approach most of us are familiar with when designing using codes.

Level II: the uncertain parameters are modelled by the statistics of the random variables (e.g., mean and standard deviation). This is the case of approximated analytical methods.

Level III: the entire joint distributions of the uncertainties are considered.

Levels I–III are all reliability-based as they ensure that the probability of failure is less than the societally accepted value.

Level IV: All uncertainties are considered in a fully-probabilistic approach, accounting for consequences of failure (e.g., losses, casualties). This is a risk-based approach.

In this book, it is essential to note that we do not suggest that an engineer should try to change the acceptable limit of the probability of failure; this is a decision to be made at the societal-political level.

REFERENCES

2.1 British Standards Institution. *BS EN 1990:2002+A1:2005: Eurocode. Basis of structural design*. London: BSI, 2005.

2.2 The Institution of Structural Engineers. *The regenerative structural engineer*. London: IStructE Ltd, 2024.

CHAPTER 3

Tools for engaging with ethical questions in design

Alix Dietzel and James Norman

3.1 INTRODUCTION

Structural engineers are waking up to the impact of their designs on the planet. The climate and biodiversity emergency contains essential ethical questions. When we started this book, we collaborated with a moral philosopher to help engineers consider these questions. This chapter raises ethical questions — specifically those central to risk-based design — which may be unfamiliar territory for readers. We hope this chapter will enable engineers to navigate ethical questions, mainly concerning climate change justice, as well as broader questions about how the conventional code approaches take account of the cost of human life. We present readers with initial tools to answer these questions when making design decisions. We focus on a few key areas, asking challenging ethical questions and thinking through the ethical implications of decision-making and the broader impacts such choices can have. As you read this chapter, try to focus not on all the things you do not yet know, but on decisions you can control within the wider context of risk and design.

3.2 ETHICAL QUESTIONS IN RISK-BASED DESIGN

Ethics is a complicated field with a rich history. Like engineering, there are many sub-fields and areas of expertise. Alix Dietzel, a co-author of this chapter, is an expert in this increasingly popular field referred to as 'climate change justice'. Scholars in this field agree on a basic premise which guides their various approaches: a focus on the moral importance of the individual as opposed to the state, community or culture. Climate justice scholars take the individual as their moral starting point because they assume that all human beings have equal moral worth and, therefore, have the right to equal moral consideration. Whenever making any decisions on what to do about climate change, they consider how individuals, no matter where they are born (or indeed when they are born because future generations have rights, too), will be affected, and consider unequal treatment of morally equal people wrong or at the very least problematic. To engage with questions of what is fair and right, climate justice scholars usually focus on what individuals worldwide deserve and how these entitlements can be distributed.

Although climate justice scholars usually assert that individuals must be the central unit of moral concern when exploring solutions to climate change, it is important to note that these scholars often prioritise different goals to ensure that individuals are the subject of equal moral concern. For example, some scholars emphasise human rights[3.1]; some discuss the importance of institutions operating fairly[3.2,3.3]. In contrast, others emphasise the importance of human capability[3.4], and still others are concerned with fair global social processes[3.3]. Alix's work, however, focuses on the importance of individuals and other non-state actors, for example, corporations, cities, groups of individual activists, or non-governmental organisations[3.5]. In this book, they are referred to under an umbrella term (non-state actors) which essentially refers to any agent that is not a state. Sometimes these actors are also called sub-state actors, networked actors or informal actors, depending on the scholar.

The post-Paris Agreement climate regime treats these actors as key to an effective policy response. From a moral standpoint, non-state actors are critically important because of their efforts to mitigate climate change, adapt to its impacts and aid those who do and will suffer from it. Most importantly, these actors can create conditions for systemic change: changing the way we do things within industries, sectors or cultures. For this reason, non-state actors have climate-related responsibility and are agents (or people, organisations) who have the potential for realising climate justice. This means that engineers and their representative institutions are agents of climate justice[3.6], particularly those who make design decisions and can make the difference between a city achieving climate goals or even a country meeting or not meeting their nationally-determined contributions under the Paris Agreement. Every piece of construction at every scale matters.

With this in mind, let us address some key ethical questions. We will explore questions focusing on wealth and the ability to pay, emissions, adapting to climate change, local and global impacts, acceptable risk of failure, and finally, reusing existing buildings. Along the way, we will draw on some of Alix's work and the work of others in the field of climate justice, but at times, we will ask questions to make you consider the role you might play in fair decision-making around climate change issues when it comes to risk.

3.2.1 Wealth and the ability to pay

Climate change will affect different countries unequally. In fact, it is estimated that less developed countries will bear 75–80% of the burdens of climate change despite contributing least to its causes[3.7]. This is for two broad reasons. First, evidence increasingly points to the fact that less developed countries face greater risk. This is in contrast with richer parts of the world. In North America, for example, climate change is projected to increase aggregate yields of rain-fed agriculture by 5–20% in the first few decades of warming[3.8]. Of course, wealthier regions will also eventually be adversely affected by climate change, but not as severely as poorer ones, because these regions have lower resilience.

The second reason less developed countries will be harder hit by climate change is that they often have additional problems, such as poverty or weak infrastructure, which create conditions of low resilience to climate change[3.9]. Less developed countries may not have the financial capacity or infrastructure necessary to combat the ill effects and will, consequently, not be able to prepare for or defend themselves against effects like flooding, droughts or rising sea levels as effectively as richer states. This raises profound questions of justice — is it fair that less developed countries must suffer from climate change if they have not caused it, and should richer countries be held responsible for adaptation costs in these countries? This approach has recently been globally accepted in the landmark COP27 agreement on Loss and Damage[3.10].

What does this have to do with risk-based design? It highlights two key issues. The first is that the wealthier a nation is, the more it can afford to build and mitigate the impact of climate change, and the poorer a nation is, the less it can afford and the more it feels the impacts of climate change. It is important to consider, when designing a building, that it is as adaptable to climate change as possible given the available resources — and this is especially important in countries that will feel effects first and hardest. When considering who might pay for such projects, appeals could be made to climate justice and the importance of protecting vulnerable populations, given their innate inability to protect themselves from climate impacts.

The second broad issue it raises is that of balancing local and global requirements and impacts. If we over-design buildings, we increase climate impact and do not meaningfully increase safety (see Part 2). From a moral standpoint, the emissions from a project will cause harm to local and global populations over time, whereas preventing the failure of a building may protect only a small number of people locally in an extreme event[3.11]. It may be worth remembering when setting out the design of a project and calculating risk, that it is not merely the immediate risks at hand that matter but the broader risks that come with emissions that rise into the atmosphere and stay there for hundreds of years, potentially risking hundreds or thousands of lives for decades and centuries to come[3.12]. It is your moral responsibility, as someone who has the relative power to change matters — for example, as someone who can influence decisions that will impact climate change — to weigh up these risks very carefully and be aware of the long-lasting impacts your decisions might have. We are not suggesting we make buildings less safe than required,

The recent Weald Action Group case shows a legal requirement to consider this, as well as a moral one.

but we are suggesting the same level of thought, or greater, should be given to these other risks.

3.2.2 Local and global impacts

Section 3.2.1 links us to the following ethical issue: all engineering projects will have a climate impact. Usually, engineers might weigh these local impacts against their immediate benefits to the local community. However, the impact of a local project will extend outwards and affect people worldwide[3.13]. It is by now very well known that the global community must act swiftly to prevent the worst effects from taking hold — such as extreme water scarcity, complete collapse of coral reefs, widespread hunger and an explosion in the number of displaced people. Responding to climate change in such a way will require rapid and systemic change across the globe. There will need to be changes in behaviour at every scale, including state, non-state and individual, and in every key industry, including transport, travel, food and infrastructure[3.14].

Climate justice scholars are acutely aware that failing to respond adequately to climate change will significantly impact human rights and worsen existing patterns of inequality across the globe. As an engineer, you can play a part in questioning existing processes within building and risk calculation, which may significantly impact people's lives worldwide. If systemic change is going to happen and happen rapidly, actors at every scale need to do their part — which is where you come in. It may be necessary to be radical and to break the usual convention in decision-making, factoring in both local and global benefits and risks of your project. You could appeal to the idea that a person's life in the local community is worth just as much as the life of a person abroad or even a person not yet born but at risk from climate change. This moral argument is based on the fundamental fact that we all have human rights, no matter where or when we are born.

3.2.3 Acceptable risk of failure

This issue raises some extremely uncomfortable truths, namely that we seem to value the lives of those locally affected by construction over those globally. Consider the risk of economic or life loss due to the failure of the building and compare it to the economic or life loss due to greenhouse gas (GHG) emissions. We may add a 'sleep at night' factor to our design to reduce the perceived risk of our building failing ('if in doubt, make it stout'), but we don't weigh against this the global risk of environmental impact. This issue is further compounded because most construction occurs in more developed countries, whereas, as previously stated, those most affected by the impact of global warming are located in less developed countries.

In the MEICON study only 6% of respondents strongly agree with the statement "The material utilisation of each structural element in my design is normally close to 1.00" and only 3% strongly agree with the statement "My clients or design team require me to minimise total embodied energy"[3.15].

Perceived risk of failure is a factor here because over-designing can mean higher emissions and, therefore, a greater risk of more deaths from climate change. We may not have realised it, but prioritising the risk of local people over the risk emissions might have on people worldwide could lead someone to think we value the lives of those locally more highly, even if this isn't a thought we have had. If you accepted less conservative (but still morally and socially acceptable) utilisation factors in a project, then you may be able to use lower emissions materials (Part 2), which would literally save lives. Of course, according to moral philosophers, there needs to be a balance between local lives and those further afield, but no human life is more important than another. A building should, therefore, be safe for those living in it without posing undue risks to other lives due to higher emissions. As an engineer, you could push for these kinds of calculations to become more mainstream, factored in more regularly and be part of conversations around risk. You could, in this way, save countless lives over your career.

Chris Wise addressed this approach in his IStructE gold medal address, where he challenged engineers to "Design to the codes … and no more"[3.16]. In Part 2, we will explore why the codes are safe enough and why we should not design beyond them.

3.2.4 Reusing existing buildings

The final and related issue is the idea of reusing existing buildings. Generally speaking, using existing structures is less GHG-intensive than completely knocking down a structure, clearing it away and starting from scratch[3.17]. However, starting from scratch is usually associated with a lower degree of commercial risk because you can control the materials used, and it may be more straightforward to calculate risk as a result. As you've probably realised from reading this chapter, such calculations prioritise lowering the commercial risk to the developer or customer and the people who end up using or occupying the building, and ignore or at least severely downplay, the climate risks to people from all over the world — including the risk of dying,

Note: "Through retrofit or reuse, the upfront embodied carbon in construction can be cut by up to 70% compared with demolishing and reconstructing, and whole life carbon can be cut by 30–75% depending on the type of building."[3.17]

but also the risk of a malnourished or difficult life because of extreme weather conditions or becoming a refugee due to climate change effects. Part 4 will explore how we can reduce and manage uncertainty on existing buildings and facilitate a move towards reuse.

As an engineer considering the risks of a project, you could play a critical part in pointing out these unfair conditions and the quiet prioritisation of certain lives over others. Of course, being that voice that makes a change is incredibly difficult. It is risky to rock the boat and demand systemic change as an actor within a bigger system that you cannot fully control. To help, we will outline two tools that may assist you in thinking through moral questions and preparing yourself for potentially uncomfortable conversations.

3.3 SOME TOOLS FOR ETHICAL QUESTIONS

Climate justice scholars have developed models of responsibility around climate change to grapple with ethical questions. The two dominant models are referred to as 'ability to pay' (the richest are the most able to pay any premiums associated with low-carbon design) and 'polluter pays' (those with the highest emissions are the most morally responsible for climate change). More recently, there have been people, including Alix, who advocate mixed models, which draw on both factors to create a kind of matrix for responsibility that weighs up both wealth and pollution[3.18]. These mixed models can capture complexities, including discerning the difference between wealthy countries that are quite green (e.g., Iceland) vs. countries that are wealthy and highly polluting (e.g., the United States) as well as being able to be 'poverty sensitive' to nations that might have high emissions but are unable to lower them as effectively due to monetary

The *Doughnut for Urban Development* includes a case study that looks at allocation of national climate budgets using similar principles[3.19].

Note: This is based on GHG emissions per country, which is what the UNFCCC bases their decision-making on, and what nationally-determined contributions under the Paris Agreement are based on[3.20].

constraints (e.g., India). Finally, these models capture the complexity of responsibility of individual agents such as engineers because they allow them to consider where they have the greatest responsibility — is it because their projects are potentially high emitting? Or because they are relatively well paid within the global context (privileged) and can, therefore, pick and choose projects that might be better for the environment or push for change in existing projects without significant risk to themselves?

Another helpful tool ethics scholars have developed is ways to think about your role as one actor in a much larger unfair system. For example, Iris Marion Young has developed a 'Social Connection Model'[3.3] that is not directly applied to climate change but works for discussions around it. She explains that all persons who participate in unjust systems, like a system that relies on GHG emissions to make a profit, are responsible for these systems because they are part of the process that causes them. They are not responsible, however, in the sense of having directed the system or intended its outcomes. For example, if we drive our cars, we are not directly aiming to harm anyone with our emissions or even upholding the system of car ownership ourselves. However, we are still contributing to the system. In these kinds of complex situations, she suggests that instead of only blaming one actor, a social connection model of responsibility allows us to call on one another to take responsibility together without blaming anyone in particular for the systems that encourage their proliferation. This does not necessarily mean that all who are involved have an equal responsibility. The power to influence the processes that produce unjust outcomes is an essential factor that distinguishes degrees of responsibility. She suggests that individuals can reason about their actions in relation to structural (i.e., built into the system or systemic) injustice along parameters of power, privilege, interest and collective ability. This kind of model may allow you to reflect on your own capacities as an engineer — what difference could you make as a just actor in an unjust system? How can you account for climate risks and minimise global impact while not compromising on the safety of the building?

3.4 CONCLUSIONS

We have raised some difficult and uncomfortable issues in this chapter, specifically about the responsibilities of engineers given their power, privilege and role in systemic issues such as risk-based design. This was not intended to arouse guilt or shame or to make you throw up your hands and give up in the face of your own sense of futility as a tiny part of a much bigger scheme of injustice. Quite the opposite; we hope you feel empowered to try to consider decisions around risk and climate change more nuancedly, even if it is harder than continuing to act as you usually would. We are no longer in a position to act in a 'business as usual' manner — it is up to all of us to change our behaviour significantly, especially in areas where we have the capacity and influence to make a radical difference.

3.5 FURTHER READING

For many engineers, ethics may be a new realm they haven't delved into previously. The following books and papers are a great place to start regarding the ethics of design and their impact on the climate more generally.

- Brock, G. *Global Justice: A Cosmopolitan Account.* Oxford: Oxford University Press, 2009.
- Hale, T. 'All Hands on Deck': the Paris Agreement and Non-state Climate Action'. *Global Environmental Politics*, 16(3), August 2016, pp12–22. Available at: https://doi.org/10.1162/GLEP_a_00362 [Accessed: January 2025].
- Heyward, C. and Roser, D. (Eds.). *Climate Justice in a Non-ideal World.* Oxford: Oxford University Press, 2016.
- Harris, P. *World Ethics and Climate Change: From International to Global Justice.* Edinburgh: Edinburgh University Press, 2010.
- Shue, H. *Climate Justice: Vulnerability and Protection.* Oxford: Oxford University Press, 2014.

REFERENCES

3.1 Caney, S. 'Climate Change, Human Rights and Moral Thresholds'. *Climate Ethics.* Oxford: Oxford University Press, 2010, pp163–180. Available at: https://global.oup.com/academic/product/climate-ethics-9780195399622?cc=gb&lang=en&# [Accessed: January 2025].

3.2 Vanderheiden, S. *Atmospheric Justice: A Political Theory of Climate Change.* Oxford: Oxford University Press, 2008.

3.3 Young, I.M. 'Responsibility and Global Justice: A Social Connection Model'. *Social Philosophy and Policy*, 23(1), 2006, pp102–130. Available at: https://www.cambridge.org/core/journals/social-philosophy-and-policy/article/abs/responsibility-and-global-justice-a-social-connection-model/9308EE478561C7CE31E1F5A8F26CBE04 [Accessed: January 2025].

3.4 Nussbaum, M. 'Beyond the Social Contract: Capabilities and Global Justice'. *Oxford Development Studies*, 32(1), 2003, pp3–18. Available at: https://doi.org/10.1080/1360081042000184093 [Accessed: January 2025].

3.5 Dietzel, A. 'Non-state Climate Change Action: Hope for Just Response to Climate Change?' *Environmental Science and*

Policy, 31, May 2022, pp128–124. Available at: https://doi.org/10.1016/j.envsci.2022.01.023 [Accessed: January 2025].

3.6 Turner, S. and Norman, J. 'Own your agency: what part will you play in the climate revolution?' *The Structural Engineer*, 99(3), March 2021, pp16–17. Available at: https://doi.org/10.56330/JQDH9427 [Accessed: January 2025].

3.7 World Bank Group. *World Development Report 2010 – Development and Climate Change*. Washington, DC: The World Bank, 2010. Available at: https://openknowledge.worldbank.org/handle/10986/4387 [Accessed: January 2025].

3.8 Intergovernmental Panel on Climate Change. 'IPCC Fourth Assessment Report: Summary for Policy Makers'. *Climate Change 2007: The Physical Science Basis*, 2007. Available at: https://www.ipcc.ch/pdf/assessment-report/ar4/syr/ar4_syr_spm.pdf [Accessed: January 2025].

3.9 Intergovernmental Panel on Climate Change. 'IPCC Third Assessment Report: Summary for Policy Makers'. *Climate Change 2001: Synthesis Report*, 2001. Available at: https://www.ipcc.ch/site/assets/uploads/2018/03/spm.pdf [Accessed: January 2025].

3.10 United Nations Climate Change. *COP27 Reaches Breakthrough Agreement on New 'Loss and Damage' Fund for Vulnerable Countries*, Press Release, 2022. Available at: https://unfccc.int/news/cop27-reaches-breakthrough-agreement-on-new-loss-and-damage-fund-for-vulnerable-countries [Accessed: January 2025].

3.11 Weald Action Group. *Horse Hill: Historic Win as Supreme Court Upholds Landmark Climate Case*. Weald Action Group, June 2024. Available at: https://www.wealdactiongroup.org.uk/2024/06/horse-hill-supreme-court-judgment/ [Accessed: January 2025].

3.12 Buis, A. *The Atmosphere: Getting a Handle on Carbon Dioxide*. NASA Science Editorial Team, 2019. Available at: https://science.nasa.gov/earth/climate-change/greenhouse-gases/the-atmosphere-getting-a-handle-on-carbon-dioxide/ [Accessed: January 2025].

3.13 The Institution of Structural Engineers. 'The second site', *The regenerative structural engineer*. London: IStructE Ltd, 2024.

3.14 Intergovernmental Panel on Climate Change. *Global Warming of 1.5°C*, 2018. Available at: https://www.ipcc.ch/sr15/ [Accessed: January 2025].

3.15 Minimising Energy in Construction. *Survey of Structural Engineering Practice Report*. University of Cambridge/University of Bath, 2018. Available at: https://static1.squarespace.com/static/58f72c9a1b631bc0c1e1b84c/t/5b85fb334fa51a1080348b19/1535507448054/MEICON+Report+Online.pdf [Accessed: January 2025].

3.16 Wise, C. 'What If Everything We Did Was Wrong?' *Building*, June 2010. Available at: https://www.building.co.uk/comment/what-if-everything-we-did-was-wrong/5000493.article [Accessed: January 2025].

3.17 The Institution of Structural Engineers. *Circular economy and reuse: guidance for designers*. London: IStructE Ltd, 2023.

3.18 Dietzel, A. *Global Justice and Climate Governance: Bridging Theory and Practice*. Edinburgh: Edinburgh University Press, 2019.

3.19 Danish Architectural Press. *The Doughnut for Urban Development: A Manual*. Copenhagen: The Danish Architectural Press, 2023.

3.20 United Nations Climate Change. *Nationally Determined Contributions (NDCs): The Paris Agreement and NDCs*. 2023. Available at: https://unfccc.int/process-and-meetings/the-paris-agreement/nationally-determined-contributions-ndcs [Accessed: January 2025].

CHAPTER 4

Brief history of reliability- and risk-based design

James Norman and Raffaele De Risi

As we conclude the first part of this book, we will look at how we got to where we are. We will explore the history of both codes (and the reasons for them) and the history of risk- and reliability-based design. This will aid us as we move into Part 2, where we explore in-depth how uncertainty is captured in the codes of practice, specifically the Eurocodes. It is worth noting that codes are not static but instead are dynamic, and change as our understanding of structures increases, our computing power improves and the approach to design changes. In Part 5, we will postulate on the future of design, which asks the question, "What next"? Rather than being against the codes, this recognises the dynamic nature of our codes.

4.1 DEVELOPMENT OF CODES

The first building code on record dates back to 1792–1750 BC. It states that:

> *"If a builder has built a house for a man, and has not made it sound, and the house falls and causes the death of the owner, that builder shall be put to death."*[4.1]

No wonder engineers, by their nature, are risk-averse!

It seems probable that it was necessary to write this statute due to many failures. Failure has often driven the development of codes.

Following the great fire of London, the London Building Act (the precursor to our current regulatory system) came into being[4.2].

Following the Ronan Point disaster, disproportionate collapse was added to building regulations, and it is now standard practice in the UK to consider this within the design stage of every building[4.3].

More recently the tragic fire at Grenfell Tower has led to a rethink of the use of cladding, which may be combustible.

However, codes don't just respond to failure; they also prevent it — and very successfully.

There are three standard approaches to code-based design.

4.2 EMPIRICAL RULES

The first design codes were not based on detailed scientific reasoning but on empirical evidence. The great fire of London led to the creation of the 1667 London Building Acts, which required buildings to be built from brick or stone rather than timber and thatch[4.4].

Empirical rules still exist today. For example, if you are looking to design a straw bale load-bearing wall (Figure 4.1), you will find that the California Building Code[4.5] provides load capacities for different wall build-ups. It also provides minimum lengths of racking walls for resisting lateral load. What it does not provide is any value of design strength.

These values have instead been calculated based on physical testing. While we might be surprised by this at first, physical testing is used regularly to provide fire ratings, fixings capacities and maximum spans for boarding material. What is less common is for these to be codified.

While more modern codes have moved away from expressing empirical solutions, they still contain a number of empirical assumptions.

Figure 4.1: Load-bearing straw bale construction designed using empirical methods

4.3 PERMISSIBLE STRESS

Following the empirical approach, engineers started to emerge who would calculate the elastic stress in materials and work out if these exceeded maximum allowable strength. This approach is known as the permissible stress approach and was used in the design of steel (for example, in BS 449[4.6] and CP113[4.7]) and concrete (CP114[4.8]). In timber design, the use of permissible stress continued into this millennium in the form of BS 5268[4.9], and permissible stress is still used in some less standard materials, such as

ISO 22156:2021[4.10], which covers the structural design of bamboo. The permissible stress approach includes the use of safety factors. Still, these are often empirical or based on judgement rather than being based on probabilistic models and are usually hidden within the permissible design strength rather than being given explicitly. The allowable stress is also limited to elastic design[4.11], and hence assumes no plasticity with all the challenges and benefits this brings.

However, permissible stress codes also have advantages. They are simple to use and do not require significant and complex calculations. While many of the codes have been superseded by Eurocodes, many engineers still prefer to use them as they are considered more conservative and are easy to apply.

4.4 RELIABILITY

Concepts about structural reliability are not new. Among others, in the 1950s, Sir Alfred Pugsley[4.12], a Professor at the University of Bristol, was already publishing concepts on safety in structural engineering[4.13]. His innovative research led to several practical implications; he left a heritage of students who continued researching, modernising and updating the field. For example, his student, Prof. Alan Garnett Davenport, presented several practical applications of structural reliability[4.14] and became North America's leading figure in reliability and wind engineering[4.15]. More recently, other academics from the University of Bristol, such as Prof. David Blockley, continued working on the concept of structural risk[4.16]. All this work echoed the effort that was progressing in North America by Prof. Allin C. Cornell[4.17], his students and distinguished colleagues. This list of names is by no means complete. However, all this effort led to the current version of the semi-probabilistic codes, hoping that reliability and risk could be directly introduced into design one day.

When these probabilistic design approaches started to appear, it was recognised that full-probabilistic design was too much to ask of a professional engineer and a more straightforward method was needed. Therefore, the load and resistance factor design (LRFD) format, also known as the semi-probabilistic method, was proposed[4.18]; this method consists of the following inequality:

$$R_d \geq S_d \quad \text{or} \quad R_n/\gamma_R \geq S_n \cdot \gamma_S \tag{4.1}$$

where R_d and S_d are the design resistance and load, respectively. R_n and S_n are the nominal resistance and action (induced by the loads); finally, γ_R and γ_S are the partial safety factors associated with the resistance and load, respectively. This inequality is the essence of the semi-probabilistic method, i.e., two numbers are compared, and no probabilities are computed. However, the assumption of specific partial safety factors implies a specific level of reliability.

The first important aspect consists of the identification of the nominal values. In structural engineering, these nominal values are known as characteristic values and correspond to a specific percentile. Specifically, the characteristic strength (Figure 4.2a) is often defined by building codes as the value for which there is a 5% probability of observing lower values; conversely, the characteristic load and resulting load effect (Figure 4.2b) is defined as the value for which there is a 5% probability of observing larger values. For variables with small variability (i.e., small coefficient of variation), a percentile equal to 50% can be used. The characteristic value is usually indicated with a subscript *k*. Since the nature of the safety factors is to provide a further level of safety, from Equation 4.1, it is possible to conclude that the design strength will be lower than the characteristic strength, and the design load will be larger than the characteristic load. Design codes provide partial safety factors that are usually larger than 1.

Figure 4.2: Distribution and characteristic values of (a) strength and (b) load

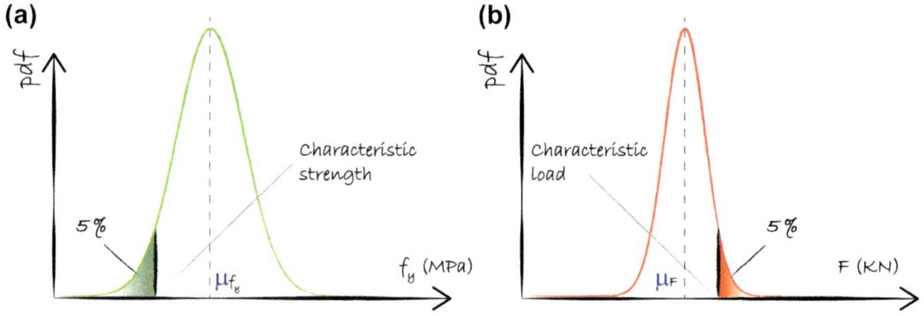

4.4.1 Limit state

The use of limit states first came about in the mid-1960s and responded to the 'factors of safety' that gave designers a false sense of security without fully understanding their basis[4.19]. A 'limit state' approach was developed where the acceptable probability of failure depended on the severity of the outcome. The concept of limit state was used in UK codes from as early as the 1970s with codes like CP110[4.20], which introduced limit state design for concrete. However, at the time, it was considered too complicated. The new BS codes BS 5950[4.21] and BS 8110[4.22] brought limit state design to both steel and concrete, respectively. It also applied the theory of plastic design in calculating section capacities (although often the analysis was elastic).

While these codes achieved the purpose of differentiating between different failure modes, it is not clear if they achieved a move from deterministic to probabilistic design, with many engineers not engaging with the probabilistic basis for the design and the assumptions that underpin the design.

With the introduction of the Eurocodes, there was a further shift to a single design philosophy (*Eurocode — Basis of Structural Design*[4.23]), which underpins all the other codes. This code covers not just the 'factors of safety' or partial factors as they are correctly known; it also makes the consequence class and expected probability of failure explicit.

However, *Eurocode — Basis of Structural Design* notes:

> *"The 'probability of failure' and its corresponding reliability index are only notional values that do not necessarily represent the actual failure rates but are used as operational values for code calibration purposes and comparison of reliability levels of structures."*

In other words, while we have moved towards a more complete statistical understanding, there remains uncertainty around this approach, and empirical data is still used to calibrate and check codes. Therefore, it is not enough to achieve an acceptable probability of failure; we also need to satisfy ourselves, as engineers, that the design is safe (Figure 4.3).

Figure 4.3: Organisation of the Eurocodes

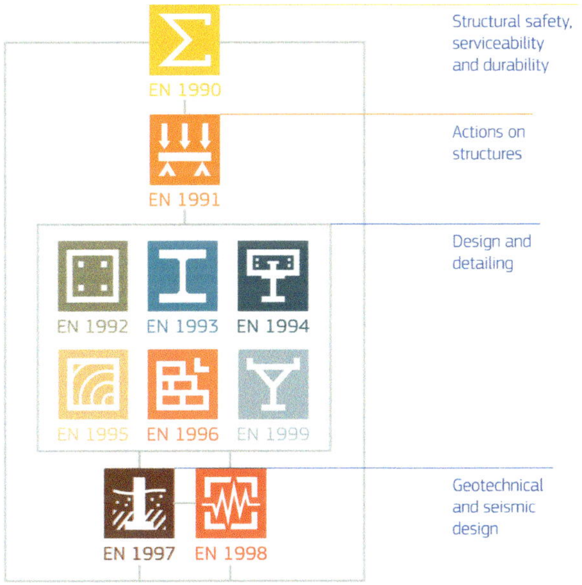

Returning to the change in codified approaches and the move to making the uncertainty more explicit, most engineers still apply the partial factors (which we still refer to as factors of safety, despite their improved scientific basis) in their approach to design and treat the design as deterministic. They don't see the full spectrum of reliability, so don't perceive the risks and opportunities.

More concerningly, engineers are often tempted to apply the same approaches to novel designs as they do to known designs, without fully considering the breadth of issues and the full suite of failure types. We aim to address both issues in this book.

REFERENCES

4.1 Frederick, M and Kuprenas, J. *101 Things I Learned in Engineering School.* New York: Three Rivers Press, 2018.

4.2 Building Talk. *How the Great Fire of London Created Building Regulations.* 2016. Available at: https://www.buildingtalk.com/blog-entry/how-the-great-fire-of-london-created-building-regulations/ [Accessed: January 2025].

4.3 The Institution of Structural Engineers. *Structural robustness and disproportionate collapse in buildings (Second edition).* London: IStructE Ltd, 2023.

4.4 Castleman, H.G. (1938), 'The Development of Construction Under the London Building Acts'. *The Structural Engineer,* 16(3), March 1938. Available at: https://www.istructe.org/journal/volumes/volume-16-(published-in-1938)/issue-3/the-development-of-construction-under-the-london-b/ [Accessed: January 2025].

4.5 California Straw Building Association. *International Residential Code Appendix AS – Strawbale Construction with Commentary,* 2018. Available at: https://www.strawbuilding.org/news/3587353 [Accessed: January 2025].

4.6 British Standards Institution. *BS 449:1932: The Use of Structural Steel in Building.* London: BSI, 1932.

4.7 British Standards Institution. BS CP113:1959: *The Structural Use of Steel in Buildings*. London: BSI, 1959.

4.8 British Standards Institution. *BS CP 114:1957: The Structural Use of Reinforced Concrete in Buildings*. London: BSI, 1957.

4.9 British Standards Institution. *BS 5268-6.1:1996: Structural Use of Timber - Code of Practice for Timber Frame Walls - Dwellings Not Exceeding Seven Storeys*. London: BSI, 1996.

4.10 ISO *22156:2021: Bamboo Structures — Bamboo Culms — Structural Design*. Geneva: ISO, 2021.

4.11 Bates, W. *Historical Structural Steelwork Handbook*. London: The British Constructional Steelwork Association Limited, 1991.

4.12 Gill, J.A. *et al.* (2009). 'Sir Alfred Pugsley OBE, FREng, FRS (1903–1998): His Legacy'. *Proceedings of the Institution of Civil Engineers — Bridge Engineering*, 162(1), March 2009, pp15–24. Available at: https://doi.org/10.1098/rsbm.1999.0028 [Accessed: January 2025].

4.13 Pugsley, A.G. 'Concepts of Safety in Structural Engineering'. *Proceedings of the Institution of Civil Engineers*, 36(5), March 1951, pp5–31. Available at: https://doi.org/10.1680/ijoti.1951.12755 [Accessed: January 2025].

4.14 Lind, N.C. and Davenport, A.G. 'Towards Practical Application of Structural Reliability Theory'. *American Concrete Institute Special Publication*, 31, January 1972, pp63–110.

4.15 Davenport, A. G. 'The Relationship of Reliability to Wind Loading'. *Journal of Wind Engineering and Industrial Aerodynamics*, 13(1–3), December 1983, pp3–27. Available at: https://doi.org/10.1016/0167-6105(83)90125-3 [Accessed: January 2025].

4.16 Blockley D. 'Structural risk: addressing the issues'. *The Structural Engineer*, 86(15), 2008, pp16–17. Available at: https://www.istructe.org/journal/volumes/volume-86-(published-in-2008)/issue-15/structural-risk-addressing-the-issues/ [Accessed: January 2025].

4.17 Cornell, C.A. 'A Probability-based Structural Code'. *American Concrete Institute Journal Proceedings*, 66(12), January 1969, pp974–985.

4.18 Galambos, T.V. 'Load and Resistance Factor Design'. *Engineering Journal*. American Institute of Steel Construction, 1982. Available at: https://www.aisc.org/globalassets/aisc/awards/tr-higgins/past-winners/load-and-resistance-factor-design.pdf [Accessed: January 2025].
4.19 Taylor, H.P.J and Burgoyne, C.J. 'Structural Design and the Eurocodes — a Historical Review'. *The Structural Engineer*, 86(20), October 2008. Available at: https://www.istructe.org/journal/volumes/volume-86-(published-in-2008)/issue-20/structural-design-and-the-eurocodes-a-historical-r/ [Accessed: January 2025].
4.20 British Standards Institution. *BS CP 110:1972: Code of Practice for Structural Use of Concrete*. London: BSI, 1972.
4.21 British Standards Institution. *BS 5950: Structural Use of Steelwork in Building (multi-part)*. London: BSI.
4.22 British Standards Institution. *BS 8110: Structural Use of Concrete (multi-part)*. London: BSI.
4.23 British Standards Institution. *BS EN 1990:2002+A1:2005: Eurocode. Basis of structural design*. London: BSI, 2005.

Part 1
Conclusions

Over the course of the last century, design has moved from an empirical to a deterministic to a probabilistic endeavour. There have been advances in understanding reliability- and risk-based design, and these advances have been applied across several specialist structures.

Our approaches need to change with the rise in awareness of the climate and biodiversity emergency and the acceleration in urgency to act. There is now a moral imperative to design a building that is safe … but no more. It is crucial to reuse what we have. It is critical to find new materials and solutions that don't just reduce harm but actually do good, and it is vital to consider the impact of climate change alongside other risks as we develop more nuanced approaches to design.

Throughout the rest of this book, we explore all these issues in greater detail.

RORY CONSTRUCTION COOPERATIVE
ALL VISITORS AND DRIVERS MUST REPORT TO THE GATEHOUSE
INCLUSIVE TOILET
AGER
MED
E CALL
NO. OF DAYS SINCE LAST ACCIDENT: 76
THIS SITE IS ACCIDENT FREE

Part 2
"...and no more"

In the near future, 90% of the carbon dioxide released over the lifetime of a new building could be from the embodied carbon. As we explored in Part 1, we need to consider both local and global risks of climate change and make our buildings as low-carbon as possible. For established materials, this means making them as materially efficient as possible.

The 90% value is based on the following logic – and while the exact number is up for debate, the fact that a large proportion is due to embodied carbon is not:

The NZCBS sets an upfront carbon limit of 430kgCO_2e/m^2 for single-family homes built in 2025[P2.1]. For the same building, it sets an energy use limit of 45kWh/m2GIA/yr. The grid runs at about 0.2kgCO_2e/kWh, so the energy use of this house in 2025 would be 9kgCO_2e/m^2. The buildings in NZCBS are all-electric, so if we assume that we have a zero-energy grid within 10 years, the total amount of operational carbon emitted by that house would be 45kgCO_2e/m^2. 430kgCO_2e/m^2 of upfront carbon emissions, and 45 kgCO_2e/m^2 of operational carbon emissions. Upfront is therefore 90% of this total.

As engineers, we also have a duty to make our designs safe. Which, where the codes exist, means designing according to the codes.

But what exactly is design? On what foundation are our codes based? How can we be confident in our analysis of the design?

Should we design beyond the codes to ensure safety, especially if we are unsure about the design, for example, using a new material?

Does making a section significantly bigger to give us a 'sleep-at-night' factor substantially change the probability of failure?

In this part of the book, we try to answer all these questions. We begin our exploration into uncertainty by introducing what reliability-based design is. We go on to explain in detail why we should, when we can, design only to the codes… and no more.

REFERENCE

P2.1 UK Net Zero Carbon Buildings Standard. *Pilot Version*. UK Net Zero Carbon Buildings, 2024.

CHAPTER 5

The purpose of design

James Norman

5.1 INTRODUCTION

As designers, we approach design with an understanding of what we need to do and what we need to prove to ensure our design is safe. Throughout our careers, we learn where to focus our attention, where the challenges may lie and which considerations are not of concern.

However, there is a problem that comes with this development. What happens if the questions we ask, the processes we go through or the assumptions we make are not the whole story? What happens when the analysis that we carry out doesn't include a governing failure case? If we focus too much on analysing and adding 'bunce' to elements that have a very small probability of failure but through 'gross error', fail to consider another failure. This chapter aims to look at our design processes and question them.

Bunce is a colloquial term used by engineers to mean extra capacity of an undefined amount.

5.2 THE PURPOSE OF DESIGN

As engineers, we are creating systems. These systems could be anything from bridges and dams to buildings or services.

These systems should fulfil a need; this could be to treat sewage or to provide a clean and safe place to care for people.

They should create value; this could be economic, or could also be much broader, for example, education, healthcare or infrastructure.

In addition, they should minimise impact or, even better, 'do no harm'. This might be fulfilled by minimising disruption to biodiversity and reducing the amount of CO_2 released. They could even go further — being regenerative in approach, enabling human and living systems to survive, thrive and co-evolve[5.1].

These systems should also have an acceptable probability of failure under normal and extreme operating conditions, and it is this final point that we shall consider further.

Of course, the acceptable probability of failure is not just about the in-use life of the system. We must consider risks during construction, repair, refurbishment and retrofitting. We need to consider dismantling, and if there is no other option, demolition. Much of this is now captured in the mindset of CDM (construction, design, maintenance), where risk must be considered at different stages.

It is not adequate to design a building or bridge in the permanent condition if there is no feasible way to construct it. As designers, we must also think through the construction stages.

Additionally, we need to consider extreme operating conditions, plus robustness, stability and disproportionate collapse. A system that offers multiple ways of working (a building with more than one load path, for example) will be more robust than a system with only one. It is not always possible to build redundancy into the system, but where we rely on a single point of failure, we must recognise this and allow it to be included in the design.

5.2.1 Calculations are only a small part of the design

Design is a complex process. It starts with a need, setting a brief, carrying out research, establishing ideas, modelling and testing these ideas, judging the outcome, and if everything is aligned, making the outcome and using it (Figure 5.1)[5.2].

Figure 5.1: The design process

While calculations are essential to the design, especially during the model, test and judge phases, design is much more than calculations. The solutions we conceive utilise the designer's professional judgement, which has developed over many years. As engineers, we use calculations to confirm that something is safe. However, we should not allow calculations to persuade us that something is safe if we have concerns.

Returning to the quote in Section 1.2:

> *"Uncertainty demands caution and caution generates waste. So what are our sources of uncertainty and how do we understand and control them?"*

Much of this part of the book will address understanding uncertainty as we improve our knowledge of uncertainty and how we characterise it. It is hoped that as we better understand the uncertainty and how it is already accounted for in the design, we can become less cautious and as a result generate less waste.

5.3 TYPES OF FAILURE

There are many different ways a structure can fail, and we will now explore some of them. It is necessary to consider how your structure might fail, and design to avoid these failure modes. Sometimes, this will require calculations that include uncertainty. At other times, the design can be done in such a way as to rule out the problem completely. This can be through material choice, geometric decisions or ensuring adequate protection. Either way, the designer needs to consider and design to avoid failure and also to understand the type of failure and its impact (which limit states they are designing for).

5.3.1 Material

When we think of structural design, specifically with steel and concrete, we often consider particular materials, ensuring they are strong or stiff enough. We check for the effects of compression, tension, bending, shear, torsion (which creates a different type of shear stress) and bearing. All of these rely on an understanding of the strength of the material.

We also check the stiffness through deflection checks and, occasionally, vibration checks.

However, material behaviour also includes a much broader spectrum of effects. We may need to design for creep (both in terms of deflection and failure), shrinkage, load duration and effects of moisture and temperature.

We may need to design for fracture or splitting, especially in brittle materials or where cracks lead to very high-stress concentrations. We should consider cyclic loading and the effects of fatigue. All of these can also lead to material failure. We will define failure more precisely in Section 6.2, but essentially failure means that the system can no longer do what it was designed for. This includes dramatic structural collapse, loss of functionality and many other aspects.

5.3.2 Degradation

Most materials degrade with time according to conditions. For example:

- Steel rusts, leading to a loss of area and, therefore, reduced capacity
- Timber, bamboo and straw can rot or be attacked by insects, leading to a loss of strength and a reduction in area
- Concrete can degrade in many ways, including through chemical changes over time (Figure 5.2)
- Earth scours as it is abraded or exposed to rain
- Plastics can change properties with extended exposure to light or due to prolonged exposure to other environmental conditions

Figure 5.2: Concrete planter with obvious signs of degradation

We must be aware of and design for the different ways our material may degrade. It is also essential to consider the types of protection, maintenance and the impact on the long-term performance of not maintaining the structure.

5.3.3 Combining materials

When we combine materials into a composite, we need to consider the behaviour of all the materials and how they work together. Steel and concrete are pretty well-matched, making a great composite in the form of reinforced concrete. But this isn't the case for all materials. Materials can change shape when subjected to stresses, changes in temperature or water content (caused by alterations in atmospheric water content and/or temperature or exposure to direct wetting). If these changes are not the same, the resulting stresses and strains in the material must be designed for, and the effects of both short-term and long-term stress variation need to be considered.

Materials can also degrade, as discussed in Section 5.3.2. The type of degradation depends on the material. Combining materials can lead to further issues relating to degradation. Examples include bi-metallic corrosion and changes in pH, which can lead to corrosion and increase the oxidation rate. It is, therefore, essential to consider these from the outset.

Composites can also change the system's load/deflection behaviour. Concrete is a brittle material under tension, but the addition of reinforcement makes the system behave ductily in bending. It is possible for brittle materials (with only an elastic load range under tensile stress) to be combined to create at least a bi-linear load-deflection response. Conversely, it is also possible that the failure becomes sudden and brittle when combining materials despite the parent materials being ductile. For example, when fibre polymer composites (FPCs) are bonded to the concrete, and the failure occurs in the bond rather than the parent material, the failure can be sudden even though the materials are under allowable tensile stress.

This change in behaviour from brittle to ductile can also be used to our advantage, as we can, by designing the bond/connection of composites, change the failure from sudden, which is undesirable in design, to progressive, which is preferred as it provides early warning of failure (for example in timber-concrete composites[5.3]).

Sometimes, we may choose to sacrifice strength for ductility. The most common example is in doubly reinforced concrete beams. We add steel reinforcement to the compression zone to ensure the steel reinforcing bars yield before the concrete fails. We could, however, achieve similar strengths by not adding this steel, but then the failure would be less ductile, and hence, we would have little warning of imminent failure.

Therefore, it is important that the composite's load/deflection response is fully understood, not just the behaviour of the parent materials.

5.3.4 Stability

Stability failure can take many forms, from the highly localised buckling of members, which is why steel sections need to be classified, to the buckling of entire frames. The collapse of large precast concrete domes is an example of minor geometric imperfections leading to complete collapse[5.4].

The consideration of stability, at some levels, is material agnostic, but as a practising engineer using a new material, we must consider it. If most of our design has been in steel and reinforced concrete, there may be blind spots that we need to be aware of.

These blind spots can include (but are not limited to) the assumption that even if not designed for, there is some tension capacity (and therefore bending), as is the case in masonry arches where when the line of thrust moves outside the arch itself the arch may collapse — it's worth noting these instances often occur **not** when the load is high (causing a significant degree of pre-compression) but when the load is low.

Another area not typically considered in steel and concrete design is thin-walled sections. Here small imperfections can lead to catastrophic collapse. Take as an example a drinks can. Stand on it, and it may well carry your load. Now, give it a small tap with an implement. The small displacement in the surface causes the can to collapse completely and quite suddenly (Figure 5.3).

So, as for all elements of design, stability is essential.

Figure 5.3: The surface of the can is gently depressed, leading to sudden and total failure

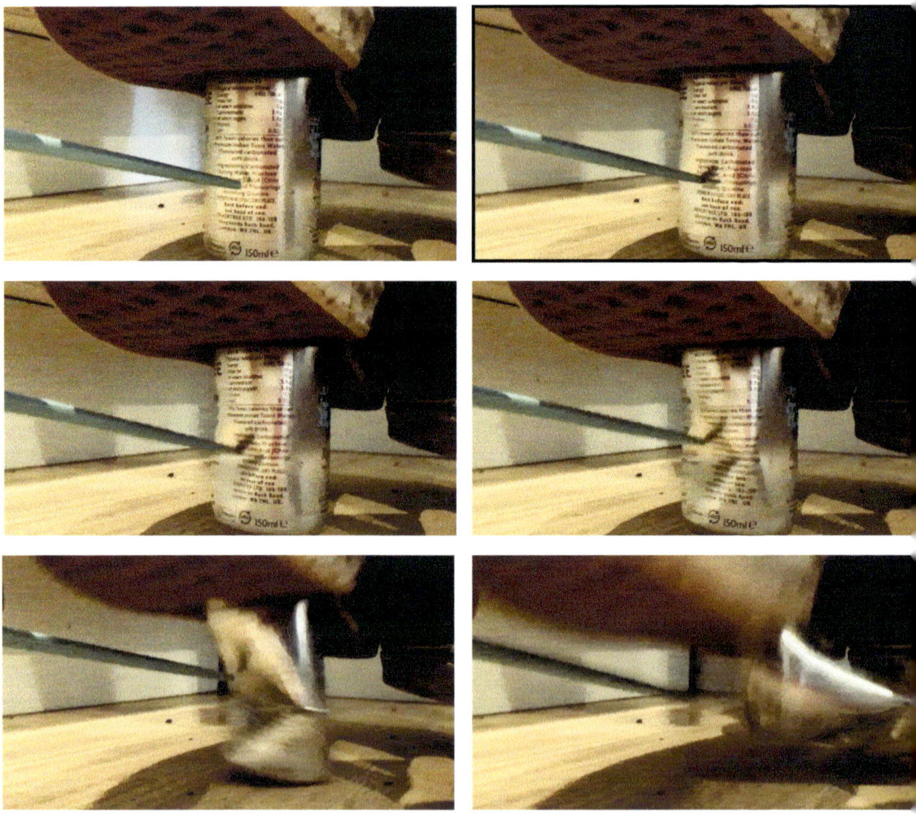

Another critical concern is not just the final stability of a frame, but also temporary conditions during construction and demolition. In both cases, there are two considerations: one, that the stability is adequate, and two, that a construction plan should be considered to ensure stability as all elements are added/removed. However, the variation in load should also be considered, primarily where the stability relies on the self-weight of the complete structure to provide it, either in the form of pre-compression within elements or to resist uplift/overturning in frames.

5.3.5 Human error

One of the leading causes of failure is human error. BS EN 1990[5.5] states that:

> *"The actual frequency of failure is significantly dependent upon human error, which are not considered in partial factor design. Thus β does not necessarily provide an indication of the actual frequency of structural failure."*

where β is the reliability index, which will be explained in more detail in Chapter 6.

Human error can be caused by failure in design (making a mistake in the calculation), detailing (drawing something that is different from the design — for example, you may have assumed a fully-fixed connection in the design but drawn a pinned connection or semi-rigid connection) and execution (what is built does not reflect what was drawn and designed) (Figure 5.4).

This is just as big an issue if a building is refurbished or repurposed. A load-bearing wall may be removed without checking the new load paths, or a space may be changed from an office to a gym without considering the change in loading. One of the most significant human errors is not to ask the question in the first place!

Figure 5.4: Masonry collapse

Human error can be reduced (but not eliminated) by adequate quality assurance. This includes checking the design, drawings and construction. Annex B4 and B5 of BS EN 1990 aim to reduce these by checking the design (B4) (Table 5.1) and the execution (B5) (Table 5.2). As designers, the checking process will be documented through the application of ISO 9001[5.6], which is then audited on a regular basis.

Table 5.1: Design supervision levels (DSL)

Design supervision levels	Characteristics	Minimum recommended requirements for checking of calculations, drawings and specifications
DSL3 (relating to RC3)	Extended supervision	Third-party checking: Checking performed by an organisation different from that which has prepared the design
DSL2 (relating to RC2)	Normal supervision	Checking by different persons from those originally responsible and in accordance with the procedure of the organisation
DSL1 (relating to RC1)	Normal supervision	Self-checking: Checking performed by the person who has prepared the design

Note: RC1–3 are described in Table 6.2 in Section 6.4

Table 5.2: Inspection levels (IL)

Inspection levels	Characteristics	Requirements
IL3 (relating to RC3)	Extended inspection	Third-party inspection
IL2 (relating to RC2)	Normal inspection	Inspection in accordance with the procedures of the organisation
IL1 (relating to RC1)	Normal inspection	Self-inspection

Note: RC1–3 are described in Table 6.2 in Section 6.4

However, human error can also be reduced by both design and communication. This may include simplifying details, using approaches that the designer or contractor understands, communicating information clearly and avoiding having multiple sources of information (especially if the contractor has not previously operated with these considerations). The designer can also make the design as robust as possible, considering alternative load paths, robustness and avoiding disproportionate collapse.

Alongside everything else, the designer should never underestimate the importance of a second opinion, and also understand that when they have been thinking about a design for a long time and the design has shifted and mutated, something which once was logical and reasonable may no longer be so. These should not be seen as failings on the part of the designer. We have all experienced the awkward review where something once rational and reasonable, through changes and iterations, has become hard to justify logically. Instead, it should be understood as an essential part of the design process to have our designs challenged to create safer designs.

5.3.6 Accidents

Accidents happen. Whether in the form of fire, an explosion, a vehicular impact or a football-sized meteor falling from the sky. The questions we need to ask for accidental design are somewhat more subtle. Generally, we are not so much concerned with failure but with the when and how of the failure.

If there is a significant fire, it is not unreasonable for the fire to lead to significant damage, maybe even partial collapse. Many buildings will need to be demolished after a fire. However, if the building collapses soon after the fire starts, while there are still occupants and the fire services are fighting the fire, this is unacceptable.

The exception is for tall buildings, which can be designed to stand in fire with compartments to prevent the spread of fire. In this case, occupants, rather than evacuated, are told to 'stay put'. The structure needs to remain sound for these buildings until the fire is either put out or burnt out (e.g., all the fuel for the fire, such as furniture, finishes and paper, has been incinerated).

For a large explosion, such as a gas explosion, the occupants of the space where the explosion occurs will sadly not survive. Still, here the question is, will the local failure lead to a disproportionate collapse, as was the case at Ronan Point[5.7]?

Accidental failures are essential to design for, but can also be challenging. We must consider how the structure performs at raised temperatures and when damaged. We need to consider whether the system can absorb the energy of large shocks, for example, from explosions. We must ensure the structure's collapse (or partial collapse) won't be disproportionate, and we need to consider the numerous facets of robustness[5.8,5.9]. The challenge for the designer with new materials is that it will be harder to acquire test data under these abnormal conditions (such as high temperatures), and it is prudent to think of ways of avoiding the question, for example, through adequate fire proofing and robust detailing and design.

Accidental failures are also critical when refurbishing a building or changing its use. We need to question whether the original design philosophy is still correct and whether the renovated building is still achieving the same approach. For example, if we assumed fire compartments in the original design, has the refurbishment maintained this, or has the design philosophy changed?

5.3.7 Extreme events

Alongside accidental actions, we also need to consider extreme events such as wind, flooding and earthquakes. It is essential to examine these in light of climate change, which alters our weather patterns and leads to a greater likelihood of extreme events.

5.4 NON-LINEARITY AND UPPER- AND LOWER-BOUND THEORY

For most designers, finite element analysis (FEA) has become an everyday tool, especially for statically indeterminate structures (frames, continuous beams, etc.) where the solution cannot be found quickly with a hand calculation. Most FEA assumes that the structure is elastic (although not all). However, when designing with the two most common materials we often design as non-linear (using plastic stress-strain distributions). For concrete, we set limits to make sure that the bars yield (hence both the need for doubly-reinforced beams for very high moments and the limit on the lever-arm Z to 0.95 of the adequate depth for very low loads); this is to ensure a ductile failure if failure were to occur. Likewise, in steel design, we use the plastic modulus for design (at least for Class 1 and 2 sections as defined in BS EN 1993-1, Section 5.5.2[5.10]) even though most of our analysis is based on linear, elastic, stress-strain relationships.

Furthermore, when we design braced frames (not portal frames), we regularly treat all connections as pinned, even when they have some rigidity (most things are neither true pins nor fully rigid, but they are somewhere in between). So, unless we create true pins, the actual design moment is not $\omega L^2/8$ but something else.

We can do this because we can apply the lower-bound theory. However, it is important to note that for the lower-bound theory to hold, there are four requirements:

- Ductility, or perhaps more strictly, deformation capacity, meaning that the yield surface is stable
- Small displacements, meaning the equilibrium equations don't change, i.e., that P-Delta effects are not significant
- Normality condition applies to the yield surface, i.e., plastic strain direction is normal to the yield surface
- Yield surface is convex[5.11]

Coupled with the lower-bound theory is the upper-bound theory. This approach is less commonly utilised but is still used in yield line analysis and portal frame design.

Both the upper-bound theory and the lower-bound theory are very powerful. They enable design to be carried out efficiently and quickly. The lower-bound theory is used on a daily basis in design offices. However, caution should be utilised as it does not always apply, especially for novel materials. Likewise, the upper-bound theory may be required for elastic design when buckling is a major concern, such as thin-walled structures, as minor geometric variations can lead to substantially different failure loads.

5.5 ACTIONS

Whenever designing a building, we need to understand what we are designing it for. We need to know the actions that our building is required to resist. For the design of buildings per EN 1990, the actions are prescribed in EN 1991-1 across seven sections. These cover general actions (EN 1991-1-1), fire (2), snow (3), wind (4), thermal (5), during execution (temporary loads) (6) and accidental actions (7)[5.12].

There has been considerable discussion recently about future flexibility and whether we should design now for extra loads to allow greater flexibility in the future[5.13]. While it is important to consider future flexibility, we need to balance this against the requirement right now to reduce the construction's carbon footprint. It is good practice to consider future uses, but an equally helpful question might be, how could we increase the capacity of our structure in the future if it were required[5.14]? And does the design make this a simple or highly complex task?

While all of this is being talked about, we occasionally come across a loading requirement that is not covered by any code of practice. Take, for example, the design of the Gorilla House at Bristol Zoo (Figure 5.5). Here, a different approach was required, which considered both the static and dynamic (impact) loading of a fully-grown gorilla. In these circumstances, we would recommend consulting the *Eurocode — Basis of Structural Design*[5.5], which states:

> *"For variable actions, the characteristic value (Q_k) shall correspond to either:*
>
> - *an upper value with an intended probability of not being exceeded or a lower value with an intended probability of being achieved, during some specific reference period*
> - *a nominal value, which may be specified in cases where a statistical distribution is not known"*

Figure 5.5: Gorilla-loading at the Gorilla House, Bristol Zoo

In other words, you need to carry out a reliability analysis. This is covered in Annex C of *Eurocode — Basis of Structural Design*, and the statistical tools required are described in Chapter 6. For variable actions, it suggests that a normal distribution can be used (although it notes that extreme value distributions would be more appropriate). Furthermore, it states that this must be included in the project documentation.

5.6 CONCLUSIONS

In this chapter, we have outlined the design process, and the role calculations play. We have explored the multitude of ways that something can fail, which we hope is a good *aide-mémoire* for an engineer, especially when faced with an unfamiliar problem. We have looked at how checking the design and construction of a building should help mitigate the risk of human error (which is not covered by the code semi-probabilistic method or any reliability design process). We explored the lower- and upper-bound theory and flagged that they cannot be applied universally but only under a limited set of conditions. And finally, we looked very briefly at loads. Underpinning all of this is the fact that rather than being determinant, everything that we design is uncertain, and in Chapter 6, we will explore this uncertainty and introduce reliability theory.

REFERENCES

5.1 The Institution of Structural Engineers. *The Regenerative Structural Engineer*. London: IStructE Ltd, 2024.

5.2 Think Up Innovations. *Describing Design as a Process*. 2018. Available at: https://thinkup.org/describing-design-as-a-process/ [Accessed: January 2025].

5.3 Mudie, J *et al.* (2020), 'Experimental Study of Moment Sharing in Multi-joist Timber-concrete Composite Floors from Zero Load up to Failure'. *Construction and Building Materials*, 225, November 2019, pp956–97. Available at: https://doi.org/10.1016/j.conbuildmat.2019.07.137 [Accessed: January 2025].

5.4 Levy, M. *et al. Why Buildings Fall Down: Why Structures Fail.* New York: W. W. Norton & Company, 1994.
5.5 British Standards Institution. *BS EN 1990:2002+A1:2005: Eurocode. Basis of structural design.* London: BSI, 2005.
5.6 *ISO 9001:2005: Quality management systems — Requirements.* Geneva: ISO, 2010.
5.7 Smale, K. 'Insight — The Ronan Point Legacy 50 Years On'. *The New Civil Engineer*, May 2018. Available at: https://www.newcivilengineer.com/latest/insight-the-ronan-point-legacy-50-years-on-16-05-2018/ [Accessed: January 2025].
5.8 The Institution of Structural Engineers. *Conceptual design of buildings.* London: IStructE Ltd, 2020.
5.9 The Institution of Structural Engineers. *Structural robustness and disproportionate collapse in buildings (Second edition).* London: IStructE Ltd, 2023.
5.10 British Standards Institution. *BS EN 1993-1-1:2005+A1:2014: Eurocode 3. Design of steel structures - General rules and rules for buildings.* London: BSI, 2005.
5.11 Calladine, C.R. *Plasticity for Engineers — Theory and Application.* Woodhead Publishing, 2000.
5.12 British Standards Institution. *BS EN 1991-1:2002-2006: Eurocode 1. Actions on structures.* London: BSI, 2006.
5.13 The Institution of Structural Engineers. *Design for Zero.* London: IStructE Ltd, 2021.
5.14 Walker, J. and Norman, J. 'Designing timber buildings for longevity'. *The Structural Engineer*, 99(11), November 2021, p4. Available at: https://doi.org/10.56330/QPOT4353 [Accessed: January 2025].

CHAPTER 6

Uncertainty in design

Raffaele De Risi and James Norman

6.1 INTRODUCTION

In Chapter 5, we unpacked the design's purpose and considered several factors that can lead to failure. In this chapter, we will start exploring how we can model uncertainty in design.

6.2 DEFINING FAILURE

To perform a structural reliability analysis, it is necessary to define a prescribed set of 'limit states', describing a condition of a system (or a component) beyond which it no longer fulfils the relevant design criteria; when this occurs, we say it has failed. In simple terms, for example, the deflection of a beam may exceed the acceptable maximum displacement (which can be governed by code recommendations, client requirements or the requirements of other building systems such as facades or roofing, which the beam needs to meet), in which case the beam fails the serviceability limit state. When the limit state is exceeded, the damage is described via the 'damage state', which provides an easy way to categorise the damage. Thus, the limit state separates the desired state from the adverse state of the system (or the component).

From an analytical point of view, the limit state is identified via a 'limit state function' $g(\mathbf{X})$. The function $g(\mathbf{X})$ is the combination of demand (load effects) and capacity. In the ultimate limit state, the demand can be the design stress, and the capacity is the design strength. In the serviceability limit state, the demand can be the expected deflection, and the capacity is the corresponding allowable deflection.

The safety margin formulation is:

$$M = C - D \qquad 6.1$$

The safety factor formulation is:

$$S = \ln\,(C/D) \qquad 6.2$$

where M and S are random variable functions of C and D. In practical applications, the safety factor is capacity divided by load effect (C/D), and the failure is attained if C/D is lower than 1.

6.3 RELIABILITY AND THE PROBABILITY OF FAILURE

Let's examine a simple structural example to explore reliability and the probability of failure. We will analyse a steel cantilever subjected to a concentrated force at the beam's free end (Figure 6.1a). The length of the beam l is equal to 6m, and the cross-section is the universal beam UB 203 × 133 × 25 (i.e., plastic modulus $W_{pl} = 258\text{cm}^3$).

When designing this beam, we assume that the design will be governed by the peak bending moment at the support. We will refer to this as the 'load effect'. In this case, the load effect consists of a single engineering demand parameter (EDP): the bending moment at the fixed end of the beam. We assume that self-weight can be neglected. We also assume that the beam will not fail in shear or any form of buckling.

The corresponding capacity for this beam can also be calculated. We will assume that the beam cross-section is Class 1. Therefore, a plastic stress distribution is assumed (Figure 6.1b). We assume the beam is somehow restrained and thus not subject to lateral torsional buckling.

For the proposed example, we have assumed that the geometry is deterministic (a set value that will not vary) and that the force F and the yielding stress are uncertain; both are assumed to be normally distributed (we will look at different types of distribution in Section 10.3.2). Specifically, the mean and standard deviation of the force are $\mu_F = 4$kN and $\sigma_F = 0.5$kN, respectively; the mean and the standard deviation of the yielding stress are $\mu_{fy} = 235$N/mm^2 and $\sigma_{fy} = 50$N/mm^2, respectively[6.1].

Figure 6.1: (a) Demand and (b) capacity models

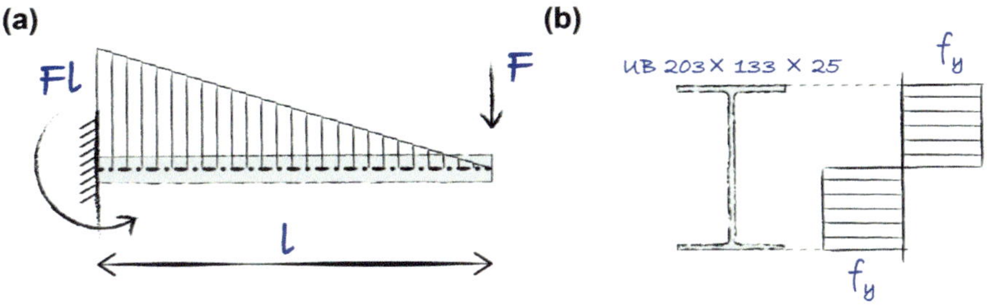

Now that we have set up the problem, let's see what happens when we run a number of analyses with the two variable inputs. We have done this using *Matlab* and *Python*, but if you are less confident with programming, you can also use Excel applying the following function:

"= NORMINV(RAND(),Mean,Standard Deviation)" 6.3

For example:

F = NORMINV(RAND(),4,0.5) = 4.478kN 6.4

And:

$$f_y = \mathrm{NORMINV(RAND(),235,50)} = 284.7\mathrm{N/mm^2} \qquad 6.5$$

Note: These numbers are randomly generated, and it is extraordinarily unlikely that you will get the same values. In fact, if you get the exact numbers, it would suggest that the random number generator in Excel is not random at all.

This can be seen diagrammatically for f_y in Figure 6.2a. If we run many simulations and represent the values using a histogram, the shape of the histogram will resemble the original probability density function (Figure 6.2b).

Figure 6.2: (a) Simulation scheme and (b) 10^5 simulations histogram

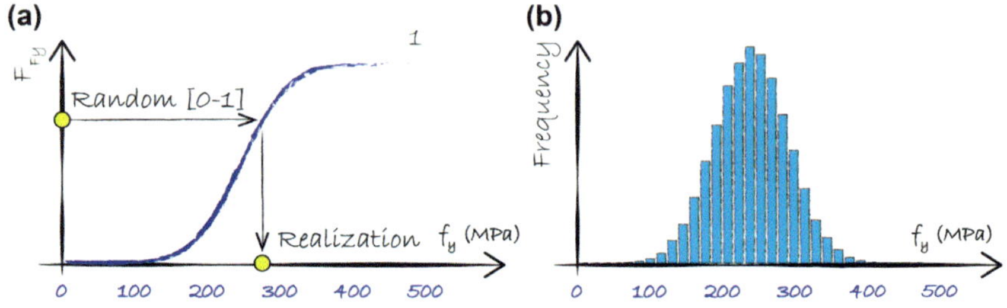

The bending moment can be calculated using a simple static equation:

$$M_{Ed} = F\,l \qquad 6.6$$

where l is taken as 6m and, in this case, we assume it is deterministic (will not vary). Of course, in reality, the length of the beam can also be modelled statistically. As the beam is a cantilever, the moment does not depend on the stiffness of the end support, which is another variable (in reality, it won't be fully fixed; it would instead have a spring stiffness depending on the connection and the stiffness of the wall).

If we were calculating the deflection, this would be another variable to consider. M_{Ed} is, therefore, the load effect.

The moment capacity of the beam is given by:

$$M_c = W_{pl} f_y \qquad 6.7$$

Again, we have set the plastic modulus as determinate and the yield stress as variable. We assume a simple bi-linear stress-strain relationship where virtually all the beam enters the plastic range. M_c is, therefore, the capacity.

Using the values given, we can see that:

$$D = M_{Ed} = 6\text{m} \times 4.478\text{kN} = 26.868\text{kN} \qquad 6.8$$

$$C = M_c = 258\text{cm}^3 \times 284.7\text{N/mm}^2 = 73.4526\text{kNm} \qquad 6.9$$

$$C/D = M_c/M_{Ed} = 2.734 \qquad 6.10$$

So, in this exact case, the beam is safe. But what about other beams with different yield stresses and maximum applied loads? Before you proceed with the example, it would be appropriate to consider whether the design will have an acceptable probability of failure.

Using a simple program, we can perform the simulation repeatedly, each time with a randomly generated F and f_y. For this straightforward analysis, you can do this process in Excel. However, a programming language like *Python* is recommended for anything more complex.

In Figure 6.3a, the result of a large number of analyses can be seen. The green dots represent cases where C/D is greater than 1, and the red dots represent cases where C/D is less than or equal to 1. In other words, the red dots are failures or unsafe, and the green dots are where the design is safe.

The probability of failure, P_F, is the total number of failures divided by the number of simulations. This is the classical frequentist definition of probability. (A detailed and mathematically rigorous background to P_F and its link to reliability is given in Appendix A.)

$$P_F = \frac{(\text{Number of times } C/D \leq 1)}{(\text{Total number of simulations})} \qquad 6.11$$

In Figure 6.3b, we plot the probability of failure, P_F, after running an increasing number of simulations (the blue dots); the moving average from these simulations is calculated and presented by the red line. As the number of simulations goes up, the value of P_F stabilises at 0.0029.

This method of analysis is referred to as the Monte Carlo method.

Figure 6.3: (a) Monte Carlo method and (b) sensitivity of P_F to the number of simulations

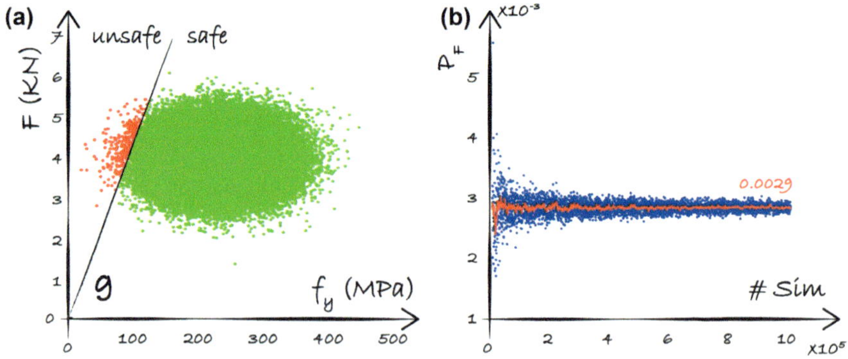

An important question to ask is how many simulations should we run? As you can see from Fig. 6.3b, the probability of failure (the red line) stabilises as the number of simulations increases. The red line becomes reasonably stable at 5×10^4 simulations.

To avoid having to do a large number of simulations (which may be computationally intensive), we can estimate the number of simulations required, $\#_{sim}$, using the following equation:

$$\#_{sim} = \frac{1 - P_F}{P_F \cdot COV^2} \qquad 6.12$$

where P_F is now the predicted order of magnitude of the probability of failure, and COV is the acceptable coefficient of variation of the estimation.

In this example, we might predict that P_F is roughly 10^{-3}, and suggest that a COV of 0.3 is acceptable. The expected number of simulations is, therefore 1.2×10^4. More in general, for a conventional realistic civil engineering problem, we may deal with a probability of failure of about 10^{-5} and an acceptable COV of 0.3. In such a case, we would need more than 1 million simulations.

It's worth noting that classical Monte Carlo methods may not be computationally efficient for a very small probability of failure (e.g., $P_F < 10^{-3}$) as the number of required simulations is inversely proportional to P_F. This means a slow convergence process with respect to the desired value of the probability of failure. To overcome such an issue, efficient simulation algorithms exist, such as the Latin Hypercube Sampling approach, the Importance Sampling technique, the Subset Simulation algorithm, the Response Surface Method, etc[6.2–6.5]. These methods either work in the input/output variables domain or adjust the simulation trajectory to have a good number of samples in the unsafe domain.

Now that we have worked out a probability of failure (0.0029), we need to ask another question: is it safe?

6.4 ACCEPTABLE PROBABILITY OF FAILURE IN RELIABILITY-BASED DESIGN

In Chapter 2, we defined mathematically the acceptable risk of failure as $P_F < P_{Ft}$ (Equation 2.2). But what is P_{Ft}, and does it vary? In this section, we investigate this in more detail.

First, we need to define the relationship between the P_F and the reliability index β. Conventionally, the reliability (or safety) index β associated with a given limit state is a function of the probability of failure:

$$\beta = -\Phi^{-1}(P_F) \qquad 6.13$$

where $\Phi(\cdot)$ is the standard normal cumulative distribution function (CDF); see Appendix A for more details. Although in Equation 6.13 the inverse of the standard CDF appears, this does not imply that uncertain parameters are distributed according to a standard normal distribution. The inversion provides a convenient one-to-one mapping between the P_F and a reliability index. It is essential to clarify that, historically, with classical methods, β was calculated first and P_F afterwards. Nowadays, P_F can be calculated directly without calculating β. However, β is still reported in the building codes as reference reliability; it is possible to argue that this is because β is easier to remember and handle with respect to P_F (e.g., $\beta = 3.09$ corresponds to $P_F = 10^{-3}$). Figure 6.4 shows the relationship between the reliability index and the probability of failure.

Note: β has a geometrical interpretation; it is the distance from 0 of a normal standard variable (presented in Appendix A). It is known that the distances (i.e., β) of 1.0, 1.96 and 3.0 correspond to an exceeding probability of 16%, 2.5% and 0.1%, respectively.

Figure 6.4: Relationship between the probability of failure and reliability index

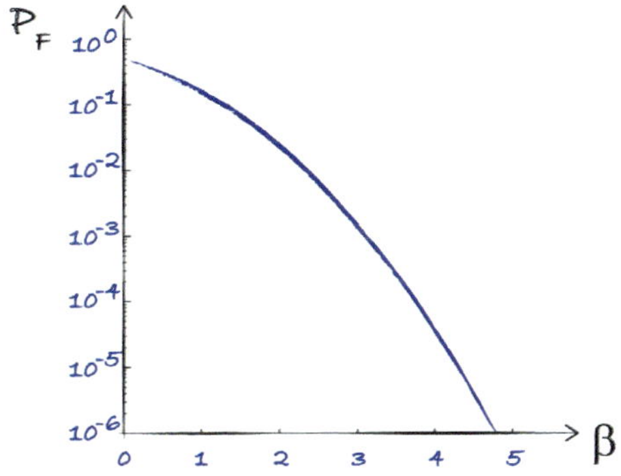

Now that we have established the relationship between the P_F and β, we can explore further which values of β we should be using in our design.

Several building codes provide partial safety factors with explicit reference to a reliability level and nominal values of strength and actions. For example, the Eurocode BS EN 1990[6.6] first provides a definition of the consequence classes (CC) (Table 6.1).

Table 6.1: Definition of consequences classes

Consequence class	Description	Examples of buildings and civil engineering works
CC3	**High** consequence for loss of human life, or economic, social or environmental consequences **very great**	Grandstands, public buildings where consequences of failure are high (e.g., a concert hall)
CC2	**Medium** consequence for loss of human life, economic, social or environmental consequences **considerable**	Residential and office buildings, public buildings where consequences of failure are medium (e.g., an office building)
CC1	**Low** consequence for loss of human life, and economic, social or environmental consequences **small or negligible**	Agricultural buildings where people do not normally enter (e.g., storage buildings), greenhouses

To the consequence classes CC1, CC2 and CC3, three reliability classes are associated, i.e., RC1, RC2 and RC3. Table 6.2 provides recommended minimum values for the reliability index associated with the different reliability classes for the ultimate limit state.

Table 6.2: Recommended minimum reliability index for ultimate limit states

Reliability class	Minimum reliability index β	
	1 year reference period	50 years reference period
RC3	5.2	4.3
RC2	4.7	3.8
RC1	4.2	3.3

Table 6.2 shows that the minimum suggested reliability index decreases with time (i.e., the probability of failure increases with time). This is because, assuming that the failure events are independent year by year, there is a higher chance of failure if the structure is exposed to the loading for a longer period.

The reliability index also depends on the considered limit state. Table 6.3 shows the reliability index for RC2 for three different limit states.

Table 6.3: Target reliability index β for Class RC2 structural members

Limit state	Target reliability index β	
	1 year reference period	50 years reference period
Ultimate	4.7	3.8
Fatigue	–	1.5 to 3.8
Serviceability (irreversible)	2.9	1.5

Similarly to Eurocodes, the American code (ASCE/SEI 2016[6.7]) first defines four risk categories, and each category provides the reliability index with a reference period of 50 years (Table 6.4).

Table 6.4: Risk categories and other structures

Use or occupancy of buildings and structures	Risk category
Buildings/structures that represent low risk to human life in the event of failure	I
Building/structures not listed in I, III and IV	II
Buildings/structures the failure of which could pose substantial risk to human life	III
Buildings/structures with potential to cause substantial economic impact and/or mass disruption of day-to-day civilian life in the event of failure Buildings/structures containing toxic or explosive substances above a certain level Buildings/structures designed as essential facilities Buildings/structures maintaining the functionality of other class IV structures	IV

Table 6.5 provides recommended minimum values for the reliability index associated with the different risk categories for the 50 years reference period[6.7].

Table 6.5: Recommended minimum reliability index for different risk categories

Basis	Risk category			
	I	II	III	IV
Failure that is not sudden and does not lead to widespread progression of damage	2.50	3.00	3.25	3.50
Failure that is either sudden or leads to widespread progression of damage	3.00	3.50	3.75	4.00
Failure that is sudden and results in widespread progression of damage	3.50	4.00	4.25	4.50

Almost in line with the mentioned building codes, the probabilistic model code (JCSS 2001[6.8]), according to ISO 2394 (2015)[6.9], suggests reliability indices for a reference period of one year, and the ultimate limit state is slightly less conservative than the previous ones. Table 6.6 shows the probabilistic model code reliability index based on the representation suggested in *Developments towards full probabilistic design codes*[6.10].

Table 6.6: Target reliability index for the ultimate limit state and 1 year reference period based on monetary optimisation

The relative cost of safety measures	Failure cost		
	Minor	Moderate	Large
Large	3.1	3.3	3.7
Normal	3.7	4.2	4.4
Small	4.2	4.4	4.7

Specifically, the β values are the results of an economic optimisation performed as a cost-benefit analysis and should not be acceptable for what concerns life safety risks. Also, such reliability values are provided as a function of failure and construction costs. For example, temporary structures and agricultural silos belong to the minor consequence class (failure-to-construction cost of about 2); residential buildings and office buildings belong to the moderate consequence class (failure-to-construction cost of about 4); hospitals, bridges, and high-rise buildings belong to the large consequence class. The relative cost of safety measures depends on the variability of the load effect and capacity[6.10]. The normal class should have a coefficient of variation between 10% and 30%. As shown in Part 4 (Reuse), since existing buildings will require more extensive efforts to increase reliability levels, the relative cost of safety measures should move from normal to large. The probabilistic model code also provides reliability index values for irreversible limit states spanning from 1.3 to 2.3. No reliability indices are provided for reversible limit states.

6.5 CONCLUSIONS

Returning to the example we set in Section 6.3, the probability of failure, P_F of 0.0029, can be converted into a reliability index of 2.76 (this can be done using tables or software tools). Assuming the beam failure may lead to medium consequence, it has a consequence class 2 (CC2 — Table 6.1) and a reliability class, RC2. As a result, for a 50-year design life, the minimum reliability index is 3.8 (Table 6.2); therefore, the beam is inadequate.

Based on your reflection at the start of the example, does this surprise you?

We suggest you try to find a steel cross-section that is commercially available and that guarantees a reliability index compatible with those indicated in Table 6.2.

Before we conclude this chapter, it is important to note:

- The reliability index does not necessarily indicate the actual frequency of structural failure
- The suggestions above are valid and can be respected at the component level (i.e., beam and columns) rather than at the system level (i.e., entire structure). This is because the link between single-element reliability and systemic reliability is complex.

In Chapter 7, we will take what we have learnt and see how it helps us to understand the basis of our design codes better.

REFERENCES

6.1 De Risi, R. *Code (produced in Matlab and Python) for the Probability Exercise in Section 6.3*. Available at: https://doi.org/10.5281/zenodo.7135313 [Accessed: January 2025].

6.2 Pinto, P.E. *et al.* 'Seismic Reliability Analysis of Structures.' *Journal of Earthquake Engineering*, 9(6), 2004, p943. Available at: https://doi.org/10.1080/13632460509350573 [Accessed: January 2025].

6.3 Ebrahimian, H. and De Risi, R. 'Seismic Reliability Assessment, Alternative Methods For'. *Encyclopedia of Earthquake Engineering*. Berlin: Springer, Berlin, 2014.

6.4 Melchers, R.E. and Beck, A.T. *Structural Reliability Analysis and Prediction*. Hoboken: John Wiley & Sons, 2018.

6.5 Nowak, A.S. and Collins, K.R. *Reliability of Structures.* London: CRC Press, 2013.
6.6 British Standards Institution. *BS EN 1990:2002+A1:2005: Eurocode. Basis of structural design.* London: BSI, 2005.
6.7 ASCE/SEI 7–16. *Minimum Design Loads and Associated Criteria for Buildings and Other Structures.* ASCE, 2017.
6.8 Joint Committee on Structural Safety. *Probabilistic Model Code.* JCSS, 2001. Available at: https://www.jcss-lc.org/jcss-probabilistic-model-code/ [Accessed: January 2025].
6.9 *ISO 2394:2015: General Principles on Reliability for Structures.* Geneva: ISO, 2010.
6.10 Vrouwenvelder, A.C.W.M. 'Developments Towards Full Probabilistic Design Codes'. *Structural Safety*, 24(2–4), April–October 2002, pp417–432. Available at: https://doi.org/10.1016/S0167-4730(02)00035-8 [Accessed: January 2025].

CHAPTER 7

Design codes

James Norman

7.1 INTRODUCTION

In Chapter 5, we explored the principles of design. In Chapter 6, we looked at reliability and the probability of failure. This chapter explores:

- How this is captured in our codes using the semi-probabilistic approach
- How we can consider different consequence classes for whole structures and parts of the structure
- What we mean by 'design life' and why risk of failure changes with time

This chapter, therefore, aims to persuade the reader that the codes are safe enough.

7.2 COMBINING THE UNCERTAINTY

As engineers, we don't just consider uncertainty once but at every step of the design process. It is helpful to understand the full breadth of uncertainty considered at the ultimate limit state (Figure 7.1)[7.1].

Figure 7.1: How uncertainty is managed at different elements of design

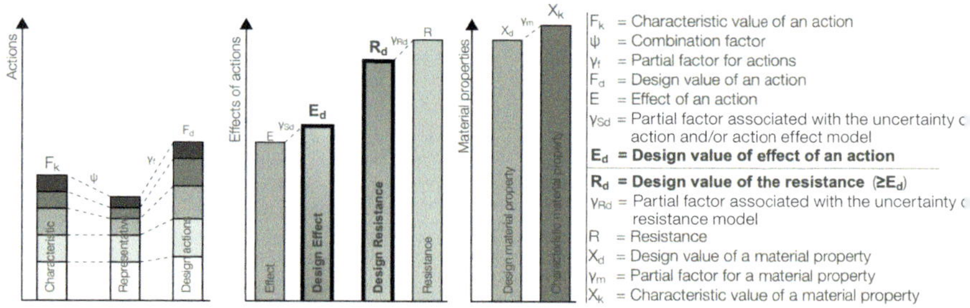

We start with an action.

We typically design not for the mean value but for the characteristic value (the 95% value; so 95% of the time, the load should be less than the value — Figure 7.2b). Actions change with time; as the building use changes, as furniture comes and goes, as people move. Hence, we design it so that there is only an anticipated 5% chance that the load will be exceeded at some point in the design's life (typically 50 years).

Figure 7.2: Distribution and characteristic values of the (a) strength and (b) load

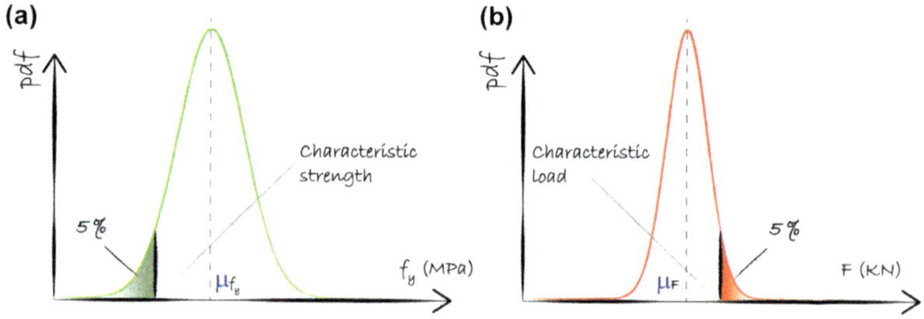

Next, we do something surprising. We apply a reduction to account for the probability of non-simultaneous actions; this could be due to combining actions (say wind and variable) or large areas (for example, when calculating column/foundation loads, we can reduce the load for multiple floors or large floor areas).

We convert the actions to design actions. We do this by applying a partial action factor (γ_f).

We then analyse the structure using these actions and apply a further partial model factor (γ_{Sd}) that changes the effect to a design effect or load effect, E_d. Note for linear analysis γ_f and γ_{Sd} are combined to be γ_F, the partial factor most engineers are accustomed to using, which accounts for action and model uncertainties and dimensional variation.

Now let's start from the other side. We have a material. As we will see in Part 3 (Classifying and testing materials), these also have a probability distribution for each characteristic, for example, tensile strength. We now take the expected 5% value, so we expect 95% of materials to be stronger (Figure 7.2a).

We then reduce the strength by dividing by a partial material factor (γ_m) to get the design strength.

Next, we reduce the design strength, which is the resistance by a partial factor associated with the uncertainty in the resistance model (γ_{Rd}) to get a design resistance or capacity, R_d. Again, for simple linear behaviour, the γ_m and γ_{Rd} are combined to give γ_M.

Finally, we compare the design effect (load effect) to the design resistance (capacity) and ensure that the effect is less than the resistance.

At each step of the way, we deal with uncertainty. Uncertainty around the action. Uncertainty around the model. Geometrical uncertainty in the model. Uncertainty around the material. Uncertainty around the resistance. And dimensional uncertainty in the resistance (calculating second moments of area, for example).

This approach is based on reliability theory but does not explicitly reference probability distributions; instead, it uses characteristic values and partial safety factors. This approach is referred to as the 'semi-probabilistic approach to design' (referred to as 'Level I' in Chapter 2)

7.3 CONSEQUENCE CLASSES AND ACCEPTABLE RISK

As alluded to in Section 6.4, we also choose a consequence class when designing a building. The consequence class is the importance of failure and considers loss of human life and the economic, social and environmental consequences. The consequence class should also consider the use of the structure. For example, in an earthquake zone, critical infrastructure failure following a severe event may cause greater loss of life than the event itself. The class affects the level of checking in design and execution (Tables 5.1 and 5.2) and the partial factors we use. Table 6.1 defines the three consequence classes and provides some examples. Most building types fall into consequence class 2, and the partial factors are those we are used to applying.

However, consequence classes don't just apply to whole buildings but can also be applied to elements within a building. We may choose to increase the partial factors for elements we deem important, such as stability or transfer members. We might design for a higher load, for example. This form of differentiation is covered in Annex B (B3.3) of *Eurocode — Basis of Structural Design*[7.2], where an increase of partial factor for actions of 1.1 is suggested for consequence class 3 (note, the increase is not 50% greater or even double, as may be practised by some engineers who apply a conservative approach, but a much more restrained 10% increase). Similarly, we may decide not to include shear links where not required in the lintels of a consequence class 2 building, even though we provide them in all primary beams regardless of whether we need them or not.

While we may not have referred to consequence classes in our design, I am sure engineers are familiar with their use, both in terms of checking and in differentiating classes across the design of a structure. As we shall see in later chapters, being confident in the application of these classes is an important part of design, whether we are using the partial factor method or a more detailed analysis of reliability.

7.4 DESIGN LIFE

Engineers need to design for the future. They need to anticipate the risks and variation in loads over time, as well as the degradation in materials. This time scale is what we call the design working life (often referred to as 'design life'); this is typically taken as 50 years for buildings.

The design life should not be confused with the use life, although it often is. The use life is how long a building is used for, and this varies greatly depending on what it is being used for. The design life is set at 50 years for most building types, although this is not always the case, as we will see shortly.

Think for a moment about the building in which you live. It may raise some questions if it is assumed it will only be good for 50 years. When was it built? How much of its 50-year life remains? What impact does this have on the value? If only one year is left, will anyone want to buy it? Or five years, or ten years, or even 20 years? But we don't ask these questions because we expect the main structure of the building in which we live to last much longer than the design life, although the building may be refurbished during this time.

In health statistics, there is an important graph, referred to as the 'bathtub' graph. It shows that you are much more likely to die at birth. As you grow, the probability of dying drops rapidly, and then rises again as we get older.

Buildings are very much the same. They are much more likely to fail in construction when they are at risk of fire (hot works), accidental impact (plant moving around), large loads (concrete being placed and not spread) and reduced stability (removing temporary propping before diaphragms are formed and stability systems are connected). Once a frame is complete, it becomes quite safe, but over time, due to a variety of factors, the likelihood of failure starts to creep up (Figure 7.3).

Figure 7.3: A prototype 'bathtub' hazard function

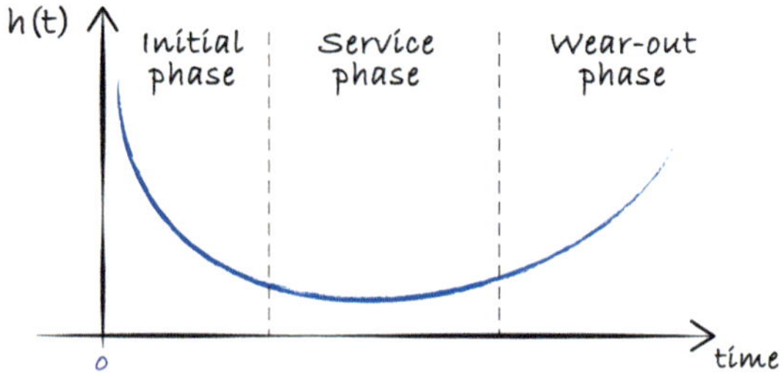

The 50-year design life then speaks to the first 50 years of life, but through regular checks and good maintenance, the building's life can be extended far beyond this value. As time progresses, the size and cost of repairs will increase, with waterproofing needing to be repaired and then replaced, and protection like galvanising and paint needing more frequent repair. Eventually, major structural renovations may be necessary, but if done in a timely manner, the building life can be extended. (More information can be found in Appendix B on the time-dependent nature of reliability.)

One final question remains — can certain items of a building be designed for a different design life? The simple answer is yes[7.2]. The most obvious example is the stability system of a building, as the cost is proportionally very small, but the impact of failure is large (and the ability to increase the capacity later in life is limited). Designing stability for a longer design life is prudent, especially considering the uncertainty around weather systems due to global warming. We may take a similar approach for different elements — either increasing the design life (often this is engineers giving a bit of extra 'bunce' for good measure — but we can actually design for this) or in certain conditions reducing the design life, for example for a temporary canopy entrance to a building which will only be used during a phase of the construction project.

Note: James was asked about this in his ICE chartership interview back in 2008 so this is certainly not a new idea.

7.5 CONCLUSIONS

In this chapter, we have seen how uncertainty is managed in design. As engineers, we can hopefully see that for loads which are codified, and for our traditional suite of materials (steel, concrete, timber, aluminium, masonry), which are also codified, the uncertainty has been accounted for through use of the semi-probabilistic method. However, as engineers, we still have an area of uncertainty in our design, the analysis of the structure and whether what we have analysed is appropriate. We will consider this thoroughly in Chapter 8.

REFERENCES

7.1 The Institution of Structural Engineers. *Design for zero*. London: IStructE Ltd, 2021.

7.2 British Standards Institution. *BS EN 1990:2002+A1:2005: Eurocode. Basis of structural design. Annex B*. London: BSI, 2005.

CHAPTER 8

Analysis

Philip Isaac and Dan Bergsagel

By now we hope you have a good understanding of the purpose of design, how we look at probabilities in terms of failure and how the risk of failure is captured by the codes to which we design.

But what about the 'model' and 'test' part in the design process outlined in Section 5.2.1? This is where engineers exercise the most judgement and manage the uncertainty in their designs, so it's important we recognise the limitations, and potential, of what we're doing and this is what we'll be covering in this chapter.

8.1 INTRODUCTION

For engineers, analysis is the bread and butter of what we do. As we become more experienced, few of us stop to consider the toils of Euler, Bernoulli, Navier, Hooke and pioneering others when formulating the equations we use to obtain bending moments or shear forces. But as things get more complicated, which analysis methods are most appropriate? How do we build confidence that the answers we're getting accurately represent real life?

Analysis is the most important step for engineers to demonstrate the suitability of their structural concepts. It's where we manage our uncertainty and build confidence in our proposal. We do this by turning real-world situations into convenient mathematical approximations that we can use to predict things like internal forces (axial forces, bending moments, shear forces) along with displacements and vibrations. We call these approximations a 'model' and will refer to them as such throughout this chapter. A model could be as simple as the linear elastic equations for a simply-supported beam in bending or as complicated as a non-linear finite element model of an entire building. Over time, the accuracy of our models has improved, aided by feedback loops through testing and a better understanding of mechanics. If we consider the example of a beam in bending, the theory used to calculate bending moments remained essentially unchanged from the Euler-Bernoulli theory of the 1750s until Timoshenko published an updated theory in the 1920s that better represented certain phenomena, such as shear deformations and rotary inertia[8.1]. But how do we decide the most appropriate 'model' to use in a certain situation? How much is there to gain from a small margin of additional accuracy? Is the extra complexity of a more representative model worth the effort instead of a simplified calculation?

Those interested in a more in-depth look at the historical development of the theories of structural analysis may wish to refer to *The History of the theory of structures: from arch analysis to computational mechanics*[8.2], which provides a detailed look at how the theories we use today were developed and by whom. The development of some of the key theories is shown in Figure 8.1.

Figure 8.1: Historical development of some key structural theories

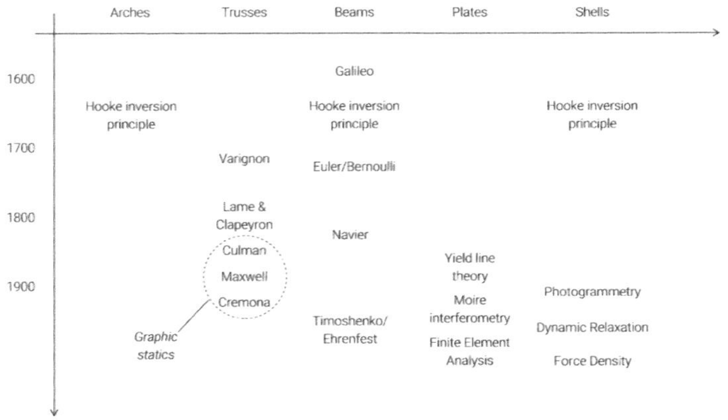

8.2 COMPOUNDING ASSUMPTIONS

The analysis process follows a set of decisions and assumptions, and each precludes certain outcomes that, if not appreciated, could lead to errors in how accurately your 'model' represents the real-world situation and behaviour you are trying to predict. Continuing with our example of a beam in bending, using the Euler-Bernoulli beam theory means you'll never be able to predict shear deformations[8.1], which could cause substantial differences in behaviour in deep beams or materials with low shear moduli (e.g., timber). This might not be an issue in most situations, but how do you know in which situation it could be a problem?

Figure 8.2 maps out some potential key decisions an engineer would make in analysis. The list is not exhaustive but summarises the types of decisions that might be relevant to the outcome.

Figure 8.2: Key decisions in analysis

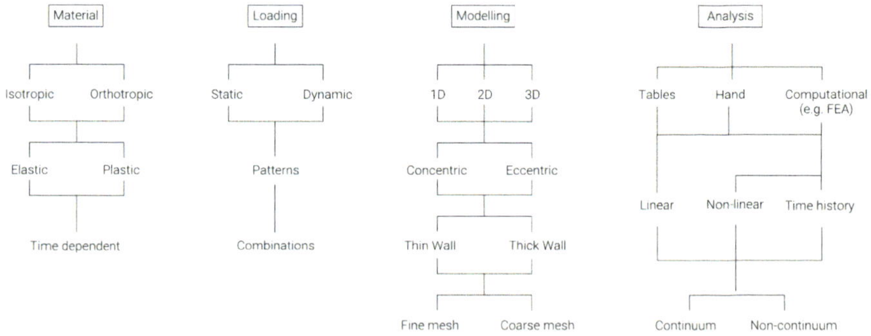

8.3 CHOOSING AN APPROPRIATE ANALYSIS METHOD

As outlined previously, the most important thing to understand at the outset is that the decisions on how you analyse a structure preclude certain structural outcomes. Fundamentally, if the model you have selected is not able to predict the relevant phenomena, then it is of little use. A starting point for this might be to envisage all the potential failure modes for the system you are analysing, and then satisfying yourself that your 'model' is able to represent these modes of failure. Examples of this include yield-line analysis in slabs where multiple patterns of yield lines are postulated and analysed, to find which has the lowest failure load, or plastic analysis in frames where it is assumed hinges form in certain locations allowing the analysis to be carried out. Similarly to yield-line analysis in slabs, classical arch analysis first requires a thrust line through the arch to be known for the analysis to be completed, which again assumes a specific distribution of forces. The aim should always be to make the simplest model possible that adequately represents the relevant phenomena, but the skill of the engineer is needed to identify what this is.

Key questions to ask at the outset of your analysis include:

- What is the purpose of the model?
- What are the driving design factors that you will need analysis information for? Strength? Deflection? Vibration? Durability?
- What is the likely failure mode or modes?
- What defines the structural system?
- What is the envisioned load path and force-resisting mechanism?
- What are the support conditions?
- How will you represent the structural system?
- Will the elements behave as cables, truss elements, beams, membranes, plates or volumes?
- Will the connections between elements be pinned, fixed, rollers or springs?
- Will the boundary constraints at the edge of the sub-assembly/at the foundations/in the ground be pinned, fixed, rollers or springs?
- Are there other elements of the building that may contribute to the behaviour that you need to consider

(e.g., infill walls can alter the natural frequency of a building which may impact its response to earthquakes)

- What other parts of the structure will you need information on to complete the analysis?
- Will you represent the whole structure, a local sub-assembly or individual elements?
- Could second-order effects or non-linearity be important?

This list applies equally to analysing existing structures as analysing new structures. However, in the case of the former the aim of the analysis is usually to determine how the structure is currently behaving, whereas in the case of the latter it's to allow us to detail the new structure.

8.4 COMMON MISTAKES

When analysing any structure there are some common mistakes that engineers can make. These include not correctly considering the material behaviour in their analysis model, inaccurately representing connections or boundary conditions and not considering non-linearity.

8.4.1 Material matters

It is important to ensure the real-world properties of the structural materials are compatible with the type of analysis you're undertaking. Many analysis 'models' that we use are independent of the material. For example, resolving forces in a truss or space frame through the method of joints is not material dependent, the material properties become important when we want to consider serviceability (deformations and/or vibration) which is dependent on material stiffness. However, some methods of analysis require material properties to be selected at the time of formulation. Finite element analysis (FEA) is an example of this as the stiffness matrix being solved requires material elastic moduli. Implicitly many of the 'models' we use assume the materials being considered follow the rules of continuum mechanics (e.g., they are homogeneous, isotropic and continuous). Linear elasticity is assumed within many methods of analysis we use (e.g., strain is proportional to stress) although plastic analysis implicitly assumes the opposite; that the material exhibits ductile behaviour. For this reason, plastic theory would not be appropriate for analysing a situation where a brittle failure mode would be expected. An example of a situation with non-continuum materials is shear in concrete, which is essentially a brittle phenomenon which may not be well-represented by continuum- or plasticity-based models.

8.4.2 Representing connections and support conditions

The fixity we assume in the connections between elements also has a major impact on the overall distribution of forces, as do the support conditions. Figures 8.3 and 8.4 demonstrate different distributions of bending moments associated with different levels of fixity at the connections (Figure 8.3) or different support conditions (Figure 8.4).

Figure 8.3: Portal frame with fixed connections, 50% rotational fixity and pinned connections

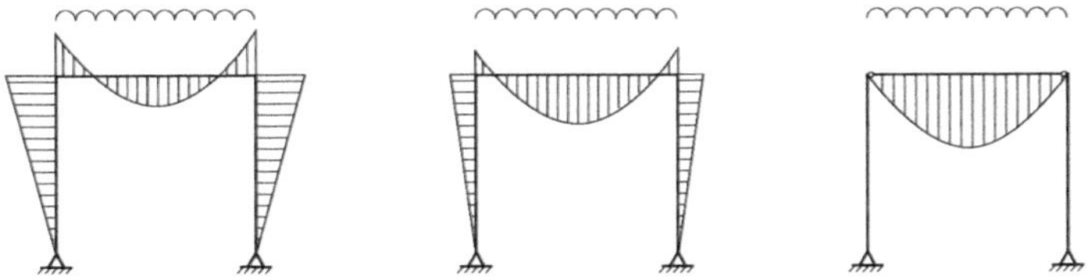

Figure 8.4: Comparing the bending moment diagrams of a frame where all three supports are fixed in position vertically (left) and where the middle support is allowed to settle under load (right)

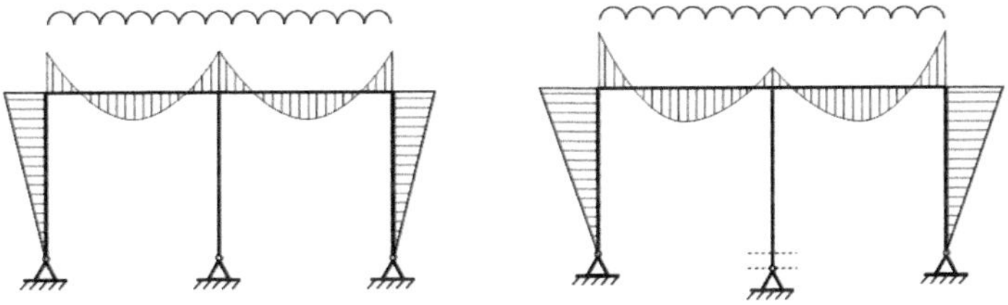

Accurately representing the support conditions requires some initial considerations to assess whether fully-rigid supports are really appropriate or if settlement or slip might be expected to occur. This can lead to what is often referred to as a 'coupled' analysis where the loads and support conditions need updating iteratively, e.g., when the load increases the settlement increases, which leads to a different distribution of forces.

8.4.3 Geometric non-linearity

Dealing with geometric non-linearity can also be challenging for some methods of analysis. For example, methods that do not model large displacements cannot assess P-Delta effects. This is where the skill of the engineer, in recognising the possible failure modes in advance, can ensure all relevant behaviours are adequately captured by the method of analysis.

8.5 WHAT ANALYSIS CAN REVEAL

It is important to recognise that analysis results rarely tell you something you do not already know about a structure's behaviour. Instead, analysis is better thought of as a tool for **confirming** and **quantifying** what you already understand. A good counter example of this is the Millennium Bridge in London, which suffered severe lateral oscillations shortly after opening[8.3]. It was later found that this was due to a phenomenon called 'lock-in'. Owing to a lack of understanding in the engineering community, the analysis methods used for the bridge were not able to account for this phenomenon, and thus it was not predicted to occur.

All analysis methods are sensitive to their inputs and therefore it is important that care should be taken here. Loading, element sizing, boundary conditions and connectivity can all play a major role in any analysis, and taking time to get this right is essential for your analysis to represent the real-world most accurately.

8.6 FINITE ELEMENT ANALYSIS

The ubiquitousness of finite element analysis (FEA) in structural engineering cannot be avoided and a brief overview of the method is outlined here, given the degree to which the method has become so popular. To find out more about the method there are many books available on the topic, such as the IStructE's *Computational engineering*[8.4].

FEA is based on continuum mechanics and as such implicitly assumes continuity in the material. This is suitable for common structural materials such as steel, or concrete in its elastic range. However, when concrete cracks, work-arounds within the formulation are required to enable FEA to represent the necessary structural behaviour. FEA works by discretising what is being analysed into smaller components — whether that be sticks or shells — and generating a stiffness matrix which defines the relationship between the connecting nodes in the discretised model. The method is particularly good for large stick models or shell models where the materials being assumed can be considered as continuums, and where the behaviour is expected to be linear with small displacements. It can also be useful in analysing structures that display geometric non-linearities, but these analyses require a form of non-linear analysis which involves the stiffness matrix being continually updated and multiple iterations assessed.

A summary of some of the advantages and disadvantages of FEA are given in Table 8.1.

Table 8.1: Advantages and disadvantages of finite element analysis

Advantages	Disadvantages
• Suitable for most commonly encountered design situations and structural materials • Able to analyse large structures • Able to run multiple iterations allowing non-linear behaviour to be assessed • Wide range of loading effects can be considered • Able to predict deformed shape • Lots of in-built functionality in most commercial packages	• Model accuracy is a function of the size of elements. However, many smaller elements means more computational power is required • Formulation is based on continuum mechanics—therefore best suited to materials and situations that exhibit this behaviour • Difficulty with representing stress concentrations unless a very detailed model is used • Over-reliance and lack of appreciation of limitations

8.6.1 Other computational methods

While FEA is undoubtedly the most common method employed in structural analysis, there are many others used in certain specific circumstances that the engineer may consider more appropriate depending on the situation to be analysed. Computational fluid dynamics (CFD) is widely used for understanding wind behaviours and interactions with buildings. Fracture mechanics-based models may be most appropriate for use in representing brittle materials. In the field of lightweight structures, dynamic relaxation and the force density method are commonly used to evaluate stresses and displacements and to undertake form-finding exercises to optimise the shape, which has a much more significant impact on the stresses than in other structural forms.

Do not underestimate the importance of identifying early which method of analysis is most suitable for your situation and do not be afraid if some research is required to explore a new method.

8.7 WHAT IS RIGHT?

It is important to recognise there are no 'right' answers from any analysis. Instead, it is better to think of it in terms of what gives us the most accurate approximation of the structural behaviour we are interested in. The term 'right' would be better thought of as 'the most certain', or framed another way, which is 'best at controlling the uncertainty'.

The only way truly to know the exact answer is to build a number of 1:1 replicas (to account for uncertainty) and test it with every conceivable loading combination (clearly an unrealistic proposition). The experience developed in the industry over hundreds of years has given us confidence in our analysis methods for estimating the accuracy of our work within familiar problems. Most engineers would feel confident in applying $\omega L^2/8$ to obtain the bending moment in a simply-supported beam, dividing this by the section modulus and obtaining a stress to compare to a material strength. But what do we do when we find ourselves outside our comfort zone?

You will rarely encounter a structural system and behaviour that has not been analysed by someone before. Therefore, if you are starting something you have no experience in, the first thing to do should be to make use of someone else's experience. Following this, if you still need further guidance, searching the literature and looking for examples of the kinds of things you're doing will usually be helpful. You may not find an example of the exact same system, but you are likely to find something similar.

Trying to recreate the results from someone else's analysis can be a good way to validate your approach before you push beyond the confines of the analysis examples you've found. Table 5.1 outlined the different levels of design supervision for different consequence classes. All but the lowest consequence class should be checked by another engineer, so knowing you are able to obtain similar results to someone else using different methods will help to build confidence.

When checking your model — or checking the work of others — some key questions you might ask could be:

- What failure mode(s) would you anticipate for the structure?
- Are the loads correct in type, magnitude and direction?
- Do the loads match the design situation and are there any special loadings that needs to be considered?
- What assumptions have been stated and what assumptions are implicit?
- Are the assumptions suitable in this circumstance?
- Are lateral forces relevant to what is being checked?
- Have the correct load combinations and factors been applied?
- What material properties have been explicitly used in the modelling, and are these supported through test data or specification?
- Do the deflections and moment diagrams match what you expected?
- Do support reactions match the loads applied?

When it comes to computational analysis it usually helps to start with a well-defined small structural system and build the size and complexity of the model incrementally. This helps to develop confidence at each step and by carrying out appropriate checks can avoid abortive work. It's important also to recognise the importance of checking, and to ensure your work is appropriately documented so that someone else can understand and check your method.

8.8 UNCERTAINTY AND CONSERVATISM

As any chef would tell you, if you put rubbish in, you'll get rubbish out. The same sentiment applies to structural analysis. Engineers often talk in terms of conservatism, but it's important to recognise that doubling the applied loads is rarely going to prevent the flaws of having carried out an inappropriate analysis finding their way into a real structure. If we think of earlier examples in this chapter, say P-Delta effects, if your analysis isn't considering large displacements then no amount of extra load to cover your uncertainty will represent this phenomenon. You, as the engineer, need to be confident the method of analysis you are using identifies the potentially critical failure modes. How you deal with uncertainty is critical to the safety of your designs. Hopefully by following some of these approaches, always acting with diligence and ensuring there is an adequate checking procedure, you can reliably analyse just about anything.

8.9 CONCLUSIONS

In this chapter we have looked at how we model our structures to enable us to analyse them. For most code-based design this area of uncertainty is where engineers have a direct impact. As a result, it is where we feel the least certain. However, the best way to manage this uncertainty is through quality assurance, not randomly adding 'bunce' and it's important to recognise that your analysis won't tell you something you don't already know. In Chapter 9 we will bring the content of Chapters 5–8 together by looking at a future case study.

REFERENCES

8.1 Öchsner, A. *Classical Beam Theories of Structural Mechanics*. Cham: Springer Nature, Switzerland AG, 2021.

8.2 Kurrer, K-E. *The history of the theory of structures: from arch analysis to computational mechanics*. Berlin: Ernst & Sohn, 2008.

8.3 Strogatz, S.H. *et al.* 'Crowd Synchrony on the Millennium Bridge'. *Nature*, 438, November 2005, pp43–44. Available at: https://doi.org/10.1038/438043a [Accessed: January 2025].

8.4 Debney, P. *Computational engineering*. London: IStructE Ltd, 2020.

CHAPTER 9

Future case study — Part 1

Raffaele De Risi

9.1 INTRODUCTION

Chapters 9, 12 and 19 of this book represent a three-part future case study. The purpose is to highlight how we might approach design in the future. It is not intended as a 'how-to' but instead to demonstrate some of the possibilities that these approaches enable.

In this first part, we present a simple example of a design for a joist of a timber roof. The design is conducted according to the Eurocodes and a reliability-based approach. Through this numerical example, we will explore the central thesis of this part; that we should design to code… and no more. We will do this by demonstrating how the codes produce designs which have an appropriate level of safety, and that by significantly increasing the section size we do not significantly reduce the probability of failure beyond the code, due to the shape of the probability function.

9.2 THE CASE STUDY

This case study focuses on a timber roof (Figure 9.1). The building is 11m long, 5m wide and 2.8m tall. The roof is constructed from 4.5m long equally-spaced joists, which span parallel to the short side of the building. Joists are fixed on a timber spreader that runs along the structure's perimeter and is supported by the rammed earth walls. On top of the joists sits a timber deck. Figure 9.1 shows further details regarding openings, some timber columns and the foundation beam, but these last components are not part of this case study.

Figure 9.1: Rammed earth house with a timber roof

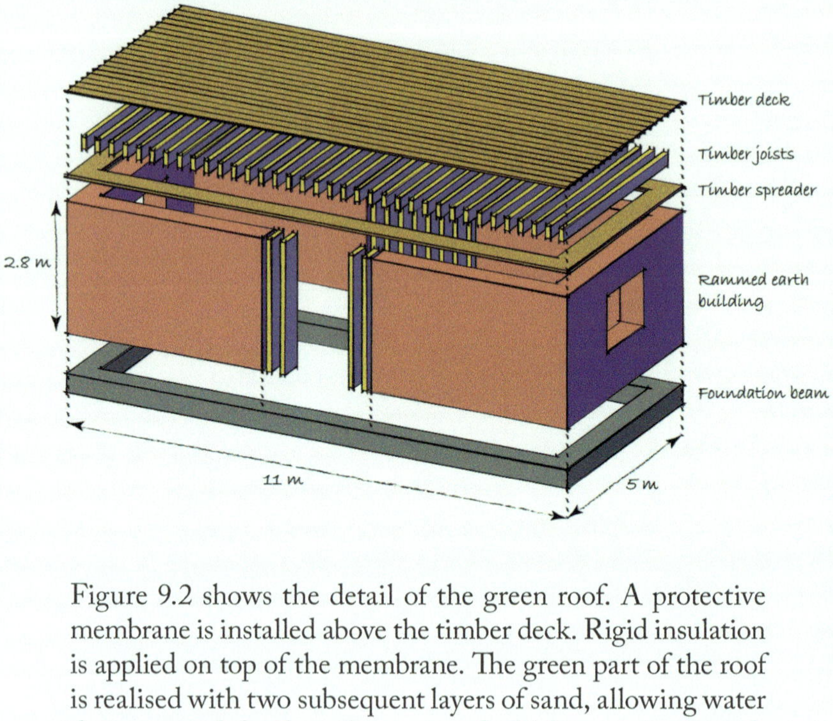

Figure 9.2 shows the detail of the green roof. A protective membrane is installed above the timber deck. Rigid insulation is applied on top of the membrane. The green part of the roof is realised with two subsequent layers of sand, allowing water drainage and fertile soil, on top of which grass is planted. The roof has a shallow incline to ensure moisture runs off the roof (not shown)[9.1].

9.2.1 Geometry

The geometry of the system cannot be considered perfect (i.e., deterministic). Therefore, uncertainties around tolerances are introduced. Specifically, the joist lengths and spacings are modelled with a normal distribution where the mean values are equal to the nominal values reported in Figures 9.1 and 9.2, and the coefficient of variation is assumed to be 0.5%. This is just an assumption; the normal distribution is chosen as it is generally good for describing the construction error. Regarding the joist cross-section and the timber deck dimensions, suggestions about the load models of the probabilistic model code[9.2] have been assumed.

Figure 9.2: Detail of the green roof

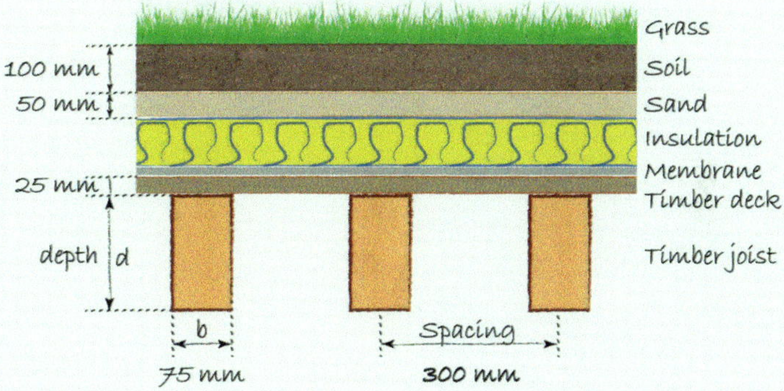

Specifically, the width and depth of the joists are modelled with a normal distribution with a mean equal to 1.05 times the nominal values and a standard deviation equal to 2mm. The timber deck thickness is modelled with a normal distribution with a mean equal to the nominal value and a standard deviation equal to 1mm. Table 9.1 summarises the assumptions on the geometry.

Table 9.1: Probabilistic description of the geometry

Parameter	Distribution	Mean	Standard deviation
Length, spacing	Normal	Nominal values	0.005 · Nominal values
Cross-section width and depth	Normal	1.05 · Nominal values	2mm
Timber deck	Normal	Nominal values	1mm

It is important to note that the assumption on cross-section width and depth should, in principle, be adopted only for the load effect model. However, such an assumption is considered consistently for the load effect and capacity models in this exercise. This will allow further discussion at the end of the case study.

9.2.2 Materials

The selected material depends on the timber availability and the design parameters. In this case study, the material properties of the timber are chosen according to standard BS EN 338[9.3], i.e., C16, C24. The adopted standard provides mean and characteristic values for the density (ρ) and stiffness properties (i.e., the elastic modulus of elasticity parallel to grain bending, E_{m}, and shear modulus, G). Unfortunately, there is no information about the mean values for the bending (f_m) and shear (f_v) strength properties; only characteristic values are provided. For reliability- and risk-based analyses, the characteristic values of the strengths are not enough. Statistics such as mean values and coefficients of variation are needed. The database of material tests compiled by Ranta-Maunus *et al.*[9.4] has been used for this purpose. The data have been enriched with further UK tests described in Ridley-Ellis *et al.*[9.5]. The availability of ready-to-use databases emphasises the need for physical testing (Chapter 11). The final database consists of mean and characteristic values and the number of performed tests for each sample collected in the databases previously mentioned. Figure 9.3 shows the weighted linear regression (red line) and its confidence interval (dotted lines) between the characteristic and the mean values of the bending strength; the weight is the number of tests. The mean value of the strength can be computed by multiplying the characteristic value provided in BS EN 338 by a normal variable having a mean of 1.85 and a standard deviation of 0.03.

Figure 9.3: Empirical relationship between the characteristic and mean tensile parallel strength

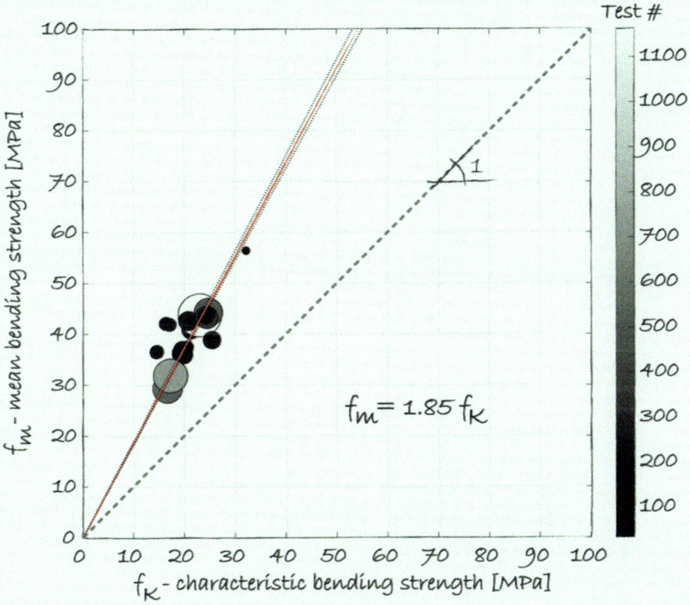

The mean value of the shear strength can be related to the mean bending strength using the following equation proposed by the probabilistic model code[9.6]:

$$f_v = 0.2 \cdot f_m^{\,0.8} \qquad 9.1$$

The probabilistic model code also provides specifications about the probability distributions to adopt for the different mechanical properties, the parameters of the distributions, and the correlation structures to assume among the timber mechanical properties. Specifically, f_m, E_m and G can be modelled with a lognormal distribution (which is covered in Part 3); the density ρ can be modelled with a normal distribution[9.6]. Table 9.2 summarises the assumptions on the material properties.

Table 9.2: Probabilistic description of the mechanical parameters

Parameter	Distribution	Mean	COV
f_m	Lognormal	BS EN 338	0.25
f_v	Equation 12.1 (perfectly correlated with f_m)		
E_m	Lognormal	BS EN 338	0.13
G	Lognormal	BS EN 338	0.13
ρ	Normal	BS EN 338	0.10

9.2.3 Limit states

As explained in Chapter 6, two limit states are conventionally considered, ultimate limit state (ULS) and serviceability limit state (SLS). ULS is associated with two potential failure modes: bending (ductile behaviour) and shear (brittle behaviour). SLS is completely calculated within the elastic behaviour.

The general limit state formulations for the ultimate flexural and shear behaviours are:

$$g_{ULS,Bending} = \frac{C_{Bending}}{D_{Bending}} \tag{9.2}$$

$$g_{ULS,Shear} = \frac{C_{Shear}}{D_{Shear}} \tag{9.3}$$

where C is the capacity and D is the load effect (or demand). The general limit state formulation for the serviceability limit state is:

$$g_{SLS} = \frac{C_{SLS}}{D_{SLS}} \tag{9.4}$$

According to what is presented in Chapter 6, these limit state functions are a simplification of the safety factor formulation, and the limit state is exceeded when $g \leq 1$.

9.2.4 Capacity

This section presents the capacity formulations for the different limit states provided. Regarding Equation 9.2, the bending capacity in terms of admissible tension can be computed according to what is suggested in Eurocode 5[9.7]:

$$C_{Bending} = k_h \cdot k_{crit} \cdot k_{sys} \cdot k_{mod} \cdot f_m \qquad 9.5$$

Regarding Equation 9.3, the shear capacity in terms of admissible tension can be computed according to what is suggested in Eurocode 5:

$$C_{Shear} = k_{sys} \cdot k_{mod} \cdot f_v \qquad 9.6$$

In previous equations, k_h is a depth factor that depends on material density and cross-section depth; its value has been computed according to Eurocode 5. The term k_{crit} is a factor accounting for lateral buckling; however, in the presence of a restraining component such as the timber deck, this can be assumed equal to 1. k_{sys} is a system strength factor that depends on how the load is redistributed among the joists; in this case study it is assumed to be 1.1 as suggested by Eurocode 5. Finally, k_{mod} is a modification factor accounting for the duration of the load and moisture content. In this exercise, service class 1 and long-term effects are considered, therefore k_{mod} is equal to 0.7. It is important to observe that all the aforementioned coefficients are uncertain. In this exercise, such uncertainty has been neglected for the sake of simplicity.

For conventional semi-probabilistic code-based structural checks, the capacity formulation provided in Equations 9.5 and 9.6 should further be divided by a partial safety factor equal to 1.3. For the purpose of this chapter, such a reduction is not needed (Chapters 6 and Chapter 7).

The capacity in terms of admissible tensions is modelled as a lognormal distribution with mean values equal to Equations 9.5 and 9.6 and a coefficient of variation equal to 10%[9.6].

Regarding Equation 9.4, the deformation capacity (C_{SLS}) can be defined as a fraction of the joist length. In this exercise, the fraction is assumed to be uniformly distributed, varying between 1/250 and 1/150. This is because, in consultation with the waterproofing supplier, it was agreed that the adopted limits would not lead to a failure in their system.

9.2.5 Load effect

The load effect on the system is computed using classical structural analysis approaches. The only difference with respect to a deterministic analysis (and given the lack of the realistic characterisation of the boundary conditions of the joist) is the assumption of both simply-supported and fully-fixed connections with a chance of occurrence of 80% and 20%, respectively. This means that results from the two schemes in terms of displacements, bending moment and shear can be weighted according to the relative probability of being the right one. Also, the two structural schemes imply that structural checks must be performed for both the middle and at the support of the joist.

Regarding Equation 9.2, the demand in terms of longitudinal tension can be computed as:

Simply-supported $$D_{Bending,\,middle} = \frac{w \cdot L^2}{8} \cdot \frac{6}{b \cdot d^2} \qquad 9.7$$

Encastré $$D_{Bending,\,middle} = \frac{w \cdot L^2}{24} \cdot \frac{6}{b \cdot d^2} \qquad 9.8$$

Encastré $$D_{Bending,\,suppport} = \frac{w \cdot L^2}{12} \cdot \frac{6}{b \cdot d^2} \qquad 9.9$$

In the case of a simply-supported beam, the load effect (D) in terms of the bending moment at the support is 0. Regarding Equation 9.3, the demand in terms of shear can be computed as:

$$D_{Shear,\,support} = 1.5\frac{w \cdot L}{2} \cdot \frac{1}{b_{mod} \cdot d} \qquad 9.10$$

where w is the linear uniform load acting on the joist, L, b and d are the length, width and depth of the joist, respectively. Finally, b_{mod} is the modified width of the joist accounting for cracking and is equal to 0.67 times b (this could also be modelled probabilistically, but we have decided, for simplicity, to use the code value). The shear load effect is reported only for the supports as it is 0 in the middle of the joist.

Regarding Equation 9.4, the load effect in terms of displacements can be computed as:

Simply-supported $$D_{SLS} = \frac{5}{384} \cdot \frac{w \cdot L^4}{E_m I} + \frac{1}{8} \cdot \frac{w \cdot L^2}{GA} \qquad 9.11$$

Fixed $$D_{SLS} = \frac{1}{384} \cdot \frac{w \cdot L^4}{E_m I} + \frac{1}{24} \cdot \frac{w \cdot L^2}{GA} \qquad 9.12$$

where E_m and G are the Young's modulus and the shear modulus of elasticity, respectively. Both elastic moduli should be divided by the term $(1 + k_{def})$, where k_{def} is a deformability factor depending on the exposure class and by the duration of the loading. For a long-term duration of the load and for a C1 class, k_{def} is equal to 0.5 (ref. 9.6). Note, Eurocode 5 suggests 0.6 for this factor but we are using the probabilistic code value. I and A are the moment of inertia and the area of the cross-section, respectively. It is important to note that the maximum displacement calculated with Equations 9.11 and 9.12 accounts for both the flexural and shear components, as the latter must be considered for timber beams.

9.2.6 Loading

The building is located in south-west England, and during this analysis we will focus on snow loading. The term ω in Equations 9.7–9.12 is obtained as the sum of the self-weight and the snow load. The self-weight is obtained by combining the geometric features and the unit weight of the materials for both the structural and non-structural components. Both self-weight and dead loads are modelled as lognormal, with a mean value equal to the nominal value and a coefficient of variation equal to 10%. The wind load is neglected here for the sake of simplicity; however, it can be treated in a similar manner.

The snow load on the flat roof is assumed to be distributed according to a lognormal distribution with a median equal to 0.087kN/m^2 and a logarithmic standard deviation equal to 0.6. This distribution is not coming from actual data as they were not available to us; however, it has a characteristic value (corresponding to the 98th percentile) equal to 0.3kN/m^2, which is consistent with the code-based characteristic value for south-west England (Eurocode 1[9.8]). For a more comprehensive method of how to collect snow data on roofs, see Lee *et al.*[9.9].

Finally, it is important to emphasise that imposed loads are also neglected for this case study to make the example as simple as possible. According to Eurocode 1, if the snow load is lower than specific values, these must be considered.

The scene is now set.

9.3 CODE-BASED STRUCTURAL CHECKS

In this section, the code-based structural check is provided for the two limit states and for the different failure modes identified previously. Code-based partial safety factors are adopted for the loading (by multiplying them by the action) and for the strength (by dividing the strength by them). Figure 9.4 shows a graphical representation of the safety factors (design capacity over design load effect; such a safety factor should not be confused with the partial safety factors) for a range of strength classes (i.e., from C14 to C40) and for a range of conventional joist cross-sections. This safety factor is consistent with the definition presented in Section 9.1 and Chapter 6. Professional engineers are more accustomed to using the utilisation factor, which is the ratio between the design load effect and the design capacity; Figure 9.5 shows the same graphical representation as in Figure 9.4 but in terms of utilisation factors.

Figure 9.4: Safety factors for multiple limit states, strength values and multiple cross-sections

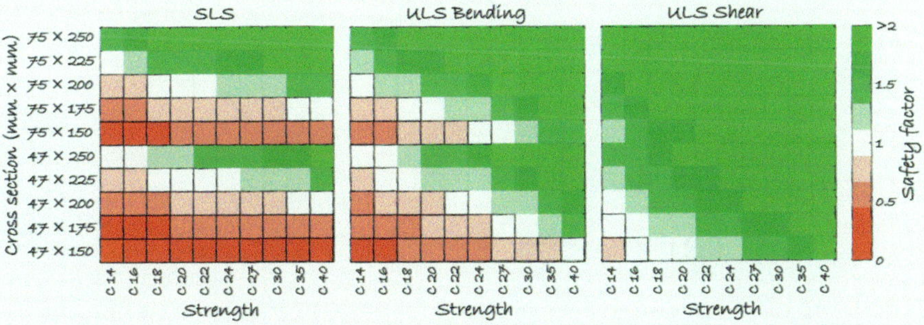

Figure 9.5: Utilisation factors for multiple limit states, strength values and multiple cross-sections

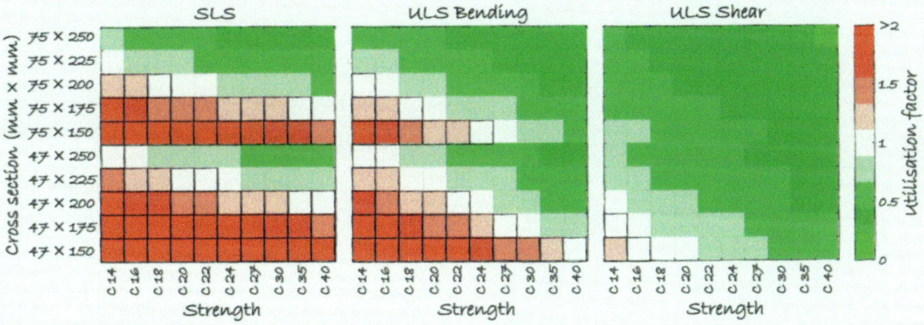

As this analysis is code-based, only the simply-supported beam is considered as the static scheme. Also, for the SLS, the limit of 1/200 is adopted for consistency with the hypothesis presented previously (i.e., the average of 1/150 and 1/250). It is possible to observe that for such a

semi-probabilistic case, the ULS is governed by bending failure, and shear is rarely a problem. Also, SLS is the critical limit state governing the design of the structural system. To simplify the interpretation of the figure, Figures 9.4 and 9.5 show squares with black borders indicating the cross-sections that are not acceptable (utilisation is greater than or equal to 1), whereas those with no borders are acceptable (utilisation less than 1). For example, a 75 × 225 C14 joist would be selected as it fulfils all criteria.

9.4 RELIABILITY-BASED DESIGN

In this section, the probability of failure for the two limit states is computed for multiple C14 timber joists having four different cross-sections. Such probability of failure is compared with the probability of failure associated with the reliability index presented in Chapter 6. Specifically, for the SLS, two limits are given: the 50-year return period and the limit corresponding to a reliability index of 2.9. For the ULS a reliability index of 4.7 is used which relates to the annual probability of failure. This is because we have defined all the probabilities over an annual period but we do not have data for a 50-year return period. Figure 9.6 shows the computed probability for joist spacing of 300mm. It can be observed that the cross-section 75 × 175 fails for both limit states. The cross-section 75 × 200 is a borderline case for the ULS, but it is not yet acceptable for the SLS. The other two cross-sections are verified for both limit states; the probability of failure for the 75 × 250 is so small that it cannot be presented graphically. Due to the nature of probability distributions, the probability of failure is reduced dramatically by increasing the cross-section sizes. However, increasing the joist spacing from 300mm to 400mm increases the probability of failures (Figure 9.7).

Figure 9.6: SLS and ULS reliability profiles for multiple C14 timber cross-sections. Spacing = 300mm

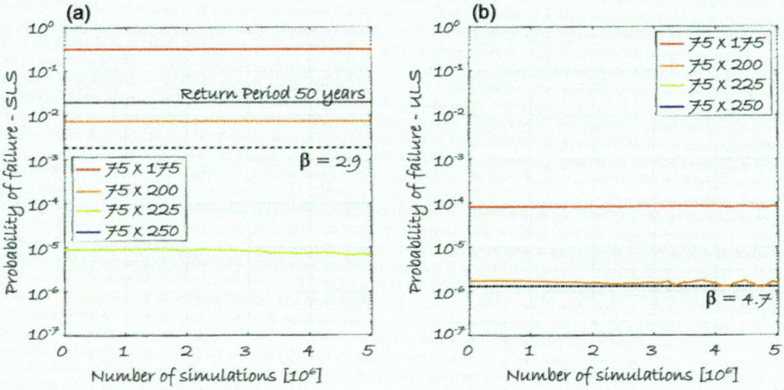

Figure 9.7: SLS and ULS reliability profiles for multiple C14 timber cross-sections. Spacing = 400mm

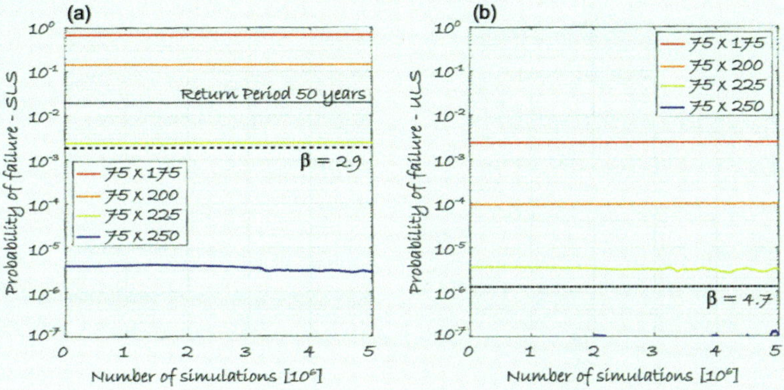

For this specific case, the reliability design is based on the check of a discrete number of cross-sections, as these are conventionally available on the market. However, for a more general approach, it is possible to fix the width of the joist cross-section and check which depth satisfies the selected reliability requirements. Figure 9.8 shows the reliability profiles for both SLS and ULS for a fixed cross-section width and multiple cross-section depths. Also, multiple material classes and joist spacings are investigated. Intersecting such profiles with the minimum reliability levels presented in Chapter 6, it is possible to identify the minimum value of the cross-section depth for adoption. For example, for a joist spacing of 300mm and a timber class C14, a minimum joist depth of about 205mm is needed to satisfy the SLS, and a minimum of about 200mm is necessary to fulfil the ULS. This is consistent with the results (orange lines) presented in Figure 9.6.

Figure 9.8: SLS and ULS reliability profiles for multiple cross-section depths for a fixed width of 75mm with various materials and joist spacings

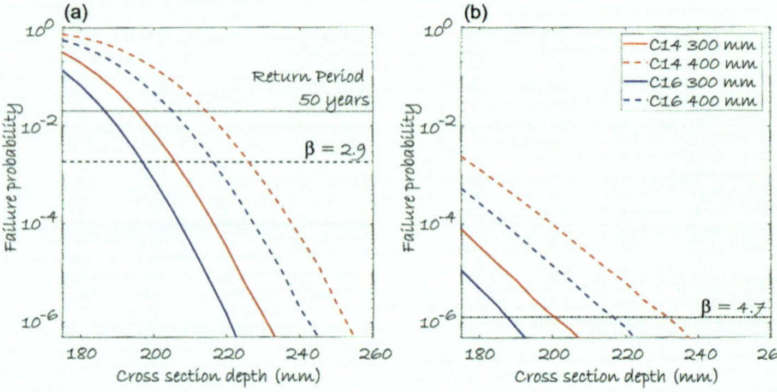

9.5 CONCLUSIONS

This simplified study of a timber joist highlights several key insights:

Firstly, that the semi-probabilistic method produces results very similar to those of the reliability-based design approach. We have played around with several parameters, but both approaches lead to similar outcomes.

Secondly, it highlights how safe our designs are. Figure 9.8 shows that the probability of failure for the 75 × 225 C14 timber joist is below 10^{-6}. For every million joists constructed, we might expect one to fail (although we also know that, even in this case, we are unlikely to see actual failure due to load sharing).

Thirdly, it highlights that once we have reached these safety levels, where the probability of failure is already very small, significantly increasing the size of elements to provide a 'sleep at night' factor does not actually lead to a significant change in the probability of failure (due to the shape of the curve) as the number is now so small that it is difficult even to imagine it.

REFERENCES

9.1 De Risi, R. *Code (produced in Matlab and Python) for implementation of Case Study in Chapter 9.* DOI: https://doi.org/10.5281/zenodo.7135337 [Accessed: January 2025].

9.2 Joint Committee on Structural Safety. *Probabilistic Model Code.* JCSS, 2001. Available at: https://www.jcss-lc.org/jcss-probabilistic-model-code/ [Accessed: January 2025].

9.3 British Standards Institution. *BS EN 338:2016. Structural Timber. Strength Class.* London: BSI. 2016.

9.4 Ranta-Maunus, A. *et al. Strength of European Timber: Part 2. Properties of Spruce and Pine Tested in Gradewood Project*, VTT, Technical Research Centre of Finland, 2011.

9.5 Ridley-Ellis, D. *Strength Grading of Timber in the UK in 2018.* Presented at *Timber 2018*, June 2018. Available at: https://www.napier.ac.uk/research-and-innovation/research-search/outputs/strength-grading-of-timber-in-the-uk-in-2018 [Accessed: January 2025].

9.6 Joint Committee on Structural Safety. *Probabilistic Model Code — Timber.* JCSS, 2006. Available at: https://www.jcss-lc.org/jcss-probabilistic-model-code/ [Accessed: January 2025].

9.7 British Standards Institution. *BS EN 1995-1-1:2004+A2:2014: Eurocode 5: Design of Timber Structures - General. Common Rules and Rules for Buildings.* London: BSI, 2014.

9.8 British Standards Institution. *NA to BS EN 1991-1-7:2006:2008 Natianal Annex to Eurocode 1. Actions on structures - General actions - Accidental actions.* London: BSI, 2008.

9.9 Lee, K.H. *et al.* 'Site-Specific Snow Load Models and Hazard Curves for Probabilistic Design'. *Natural Hazards Review*, 6(3), August 2005. Available at: https://doi.org/10.1061/(ASCE)1527-6988(2005)6:3(109) [Accessed: January 2025].

Part 2 Conclusion – "...and no more"

In this part, we have unpacked the purpose of design and the different forms that failure can take. We have explored how we propagate uncertainty using the Monte Carlo method and how this relates to the reliability we aim for. We have also looked at how the codes work and how they use reliability. We have considered de-risking design and construction by applying the level of checking necessary for the building's consequence class. Finally, we have explored analysis in more depth, considering the limitations of analysis and how we can check our analysis to verify it and reduce the risk of mistakes.

The purpose of this is to highlight that the codes, far from being a bare minimum which, as engineers, we should always try to **far** exceed, have, in fact, already accounted for a variety of unseen situations (but not a gross error), and that, therefore if we have carried out the design and have had it checked by a competent person (to reduce the chance of gross error) then we should design to the code ...and no more.

If we have concerns about our design (maybe there is something novel or unfamiliar included), significantly increasing the factor of safety won't make it any safer in meaningful terms. It may hide a gross error that won't be resolved by adding a bit of 'bunce'!

As engineers, we have a duty to make our designs safe. That means designing to code. We also have a duty to make our planet safe, and that means designing to code and no more.

In Part 3, we consider what happens when there is no code.

What if we are designing with a new material or one whose codes haven't been fully characterised? How should we proceed, and how do we characterise materials in this case?

Part 3 Classifying and testing materials

In *The regenerative structural engineer*[P3.1] an approach of 'continuous, place-based design' is presented.

This approach encourages engineers to reimagine design by starting with the question of what is available and what is abundant, that we can use to enable human and living systems to survive, thrive and co-evolve.

Often, the answer to these questions leads to more questions, such as, what if the solution requires a design that is not covered by our codes of practice? What if we are using timber that hasn't been graded or agricultural waste that is untested? What if we have a new type of concrete which needs classifying?

In Part 3, we explore these questions in more detail by considering how we can classify materials. For those properties that require a probability distribution, which distributions are suitable, and which should we use?

Of course, to answer these questions, we will need to do some physical testing. But what type of testing is necessary? What information is already available? Should we carry out the tests ourselves or work with a testing house? In the second half of this part, we will explore all these questions and more.

REFERENCE

P3.1 The Institution of Structural Engineers. *The regenerative structural engineer*. London: IStructE Ltd, 2024.

CHAPTER 10

Classifying material properties

Dan Maskell and Raffaele De Risi

10.1 INTRODUCTION

In this chapter, we will be looking at questions we should be asking about materials. We will start by looking at how materials are categorised, focusing on innovations through the lens of environmental impact. We will also look at how we can model the uncertainties associated with material properties, exploring various properties we may need to consider. We will examine how we can utilise this information to select a material and provide a list of further questions to consider.

From discussions in Part 1 and the climate and biodiversity emergency all around us, it is clear that we must change our approach to using new materials. As such, this chapter may be quickly out-of-date if we simply provide a list of properties or recommended materials. Instead, this chapter attempts to challenge our approach to materials.

10.2 THE ENVIRONMENTAL IMPACT OF MATERIALS

Structural engineering is inherently a tactile discipline, where engineers must transfer their understanding of materials' behaviour into the built physical form. It is typical to reduce a material behaviour into individual and often discrete material properties that, while making it easier to design with, can lead to a dissociation of how any material may respond under any given action. Material properties are deeply entwined with structural mechanics and how we design and construct structural systems. As highlighted in Chapter 8, many methods of analysis make fundamental assumptions about materials and their related properties and behaviour, and these assumptions may not always be appropriate, especially for non-conventional materials. Therefore, when approaching any new design, the suitability and fundamentals of the material behaviour should be considered.

The need to consider material properties and their appropriate use has never been more important than when considering the environmental impact and climate and biodiversity emergency. The environmental impact of construction is quantified by the whole life cycle impact of a construction process, in use and end of life. Typical measures of environmental impact used throughout the construction industry include the measure of energy and carbon, and while there are other measures of impact, collectively energy and carbon will be referred to herein as the 'environmental impact' as they are currently the best understood, and what engineers are most able to calculate. The 'embodied environmental impact' is related to creation and maintenance of buildings and is largely attributed to the manufacture of building materials. The 'operational environmental impact' is the

Note: Work has started on quantifying biodiversity impact in a similar way[10.1,10.2].

impact from running systems within the building, typically to control the indoor environment.

Historically, the embodied carbon accounts for only about 20% of the overall building impact[10.3], which has resulted in a legislative focus on improving operational performance. While this has beneficially reduced the impact of buildings, it has made the embodied carbon much more significant. The embodied carbon from buildings constructed today accounts for approximately 50% of the full-life cycle impact and the structure alone can comprise around 30–50% of the embodied impact[10.3].

Some of the most influential factors in reducing the environmental impact of buildings are through material choice, design lifespan, efficient design and the increase in material/system reuse to approach circularity. For efficient design, elements of buildings must be appropriately optimised to ensure they will not cause structural failure but also do not waste material. In design, once the overall concept design for a building is complete, there may be multiple designs for the structural elements that satisfy the functional requirements. While it is key to ensure that each of the functional requirements is satisfactorily met, significant overspecification will lead to excess material use. Ideally, all the different requirements will be met and no more, as outlined in Part 2. This, however, assumes that there is a large variety of materials to be used within construction where properties can be (and, more importantly, are) tailored.

10.2.1 Processing

Almost any material used within construction undergoes an element of processing typically to achieve desirable behaviour; either *in situ*, during construction, off-site during manufacture and often a combination of all scenarios. Industrialised materials are commonly categorised as those that have undergone significant processing often within a factory. Beyond the benefits of economies of scale, the factory setting often results in replicability and quality control. Replicability reduces variation in materials' behaviour and properties, with quality control checking that specific requirements are achieved. Quality control can consider multiple parameters to assess the performance directly or indirectly. These two aspects are key to reducing the variability of material properties, reducing uncertainty and increasing confidence when designing and building with materials. This is arguably beneficial and desirable for any material, especially those used for structural applications where insurance of safety is a key factor. This has resulted in conventional materials used within structures being those that are 'industrialised' and highly processed.

However, the processing required to achieve this standardisation in materials often carries the burden of higher environmental impact. Steel and cement, two of the most commonly-used materials for structures, are also characterised by an energy-intensive activity, with both taking place at high temperatures (approximately 1,400°C for cement and 2,000°C for steel). These high temperatures mean a high embodied energy and an associated embodied global warming potential (often referred to as 'embodied carbon') from the production of this energy. Furthermore, the process can directly release carbon dioxide into the environment, such as through the sintering process of cement clinker where limestone decarbonates ($CaCO_3$ –> $CaO + CO_2$).

Understanding and investigating the processing and, more broadly, the manufacturing methods can make it more environmentally efficient. Traditionally this drive for efficiency has been focused on making the process more economical, but life cycle analysis (LCA) can indicate the significance of each stage of the material and consider how they can be optimised for reduced environmental impact.

10.2.2 Categorising (to help understand the unknown)

There is a huge variety of materials available to be used within any structural system, and arguably even more so for building design more generally, especially when we consider the wider context of infrastructure for our built environment. Describing what constitutes an engineering material can be complex, and this range of materials can be categorised and grouped together in numerous ways (Figure 10.1) often based on the intended application, their composition, origin or historical use.

An example of this complexity can be demonstrated by considering concrete. Concrete is one of the most common materials in construction. Still, it can be considered a man-made composite material typically consisting of various natural aggregates, cement, and water. Concrete can be used both *in situ* on-site and prefabricated off-site and made into a variety of forms from masonry units to columns and beams, to elements such as walls and slabs.

Figure 10.1: Categorising materials

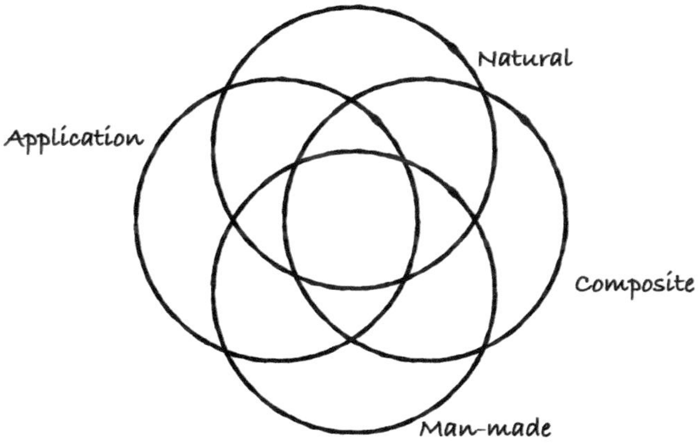

As discussed, processing can result in consistency in performance but is often associated with an environmental burden. There is no defined definition of 'significant processing', and while through LCA we can quantify the impact of manufacturing materials, there are also no threshold values and, therefore, it is intrinsically subjective. This extends to the definition of 'industrialised' materials, with the 'non-industrialised' often being described as 'natural', 'low carbon', or 'non-conventional materials'.

While within structural applications we may typically focus on the properties and structural mechanics of concrete and steel due to their abundance of use, there are many other materials that we could consider. Each material has differing and unique properties, which offers advantages and disadvantages over conventional materials.

10.2.3 Innovations with materials — historical context

Historically, construction was limited to small-scale dwellings with the intention of providing basic protection from the external environment. Initially, natural geological formations, such as caves, were primarily utilised. There are signs of permanent buildings constructed from local materials as far back as 45,000 years ago in Indonesia. Natural materials such as clay that can be easily moulded by hand and forgeable items, including leaves and branches, were then used for simple shelters. This progressed with the use of timber and stone for construction as humans developed tools and methods. Early forms of concrete-like materials have been discovered from as early as 6,500 BC, and although it was often used later in the Roman Empire, modern concrete can be attributed to 1824 with the patent for Portland cement. Fired bricks first appeared in Neolithic China around 4,000 BC but became prevalent in Europe in the 12th century and increased during the Industrial Revolution. Within the 19th century, along with Portland cement-based concrete, reinforced concrete and steel framed construction were first used, becoming conventional in the 20th and 21st century. The relative speed, ease of mass production of these materials, ability to form/fabricate various shapes/profiles, relatively low cost and compatible parameters have led to dominant steel and concrete. This prevalence resulted in further development and, thus, greater understanding, which then propagates their use.

Although other materials have seen development and implementation, including structural glass and fibre polymer composites, these have not had the same impact within the industry. Arguably, the dominance of a few materials has resulted in a lack of innovation within the construction industry with respect to material selection, which can in part be attributed to our approach to the materials developed during the Industrial Revolution.

We now have several hundred years of experience with these materials, improving the manufacturing and understanding of the structural mechanics and construction methods, allowing for a degree of optimisation. Historically, this has led to research and development focused on these limited materials to enhance their mechanical properties and make them cheaper, further propagating their use within the industry. This has enabled greater confidence in their material properties as well as methods of design, which in turn has further driven their use and, therefore, development. However, counter to beneficial confidence and improvement of conventional materials, there is arguably an over-reliance on these conventional materials and an aversion to non-conventional materials. This has allowed for over-specification, where the material performance far exceeds actual requirements, enabled by the relatively low cost of material and confidence in the design. While intrinsically this isn't a criticism, with the need to consider the impact of materials used for construction, alternative and innovative materials are required.

There are numerous challenges to using innovative materials, which broadly include political, regulatory and industrial challenges beyond just a lack of awareness and understanding, as well as a general resistance to change. While many governments worldwide have recognised the need for sustainable construction, this is often focused on the operational performance of the building rather than the embodied impact of the building and the materials. There are very few incentives to innovate with non-conventional materials in an industry where materials are relatively cheap, and any value-added properties are negated by a perception of risk. One of the greatest strengths of the industry is its size, which can lead to significant positive impacts, especially with respect to the environment, if we were to see a shift in material usage. However, this size propagates cost competitiveness through economies of scale, but also in complexities of the regulatory framework, which combined leads to an arguably risk-averse industry that fails to innovate in the use of materials.

Combined, the current and historical context of the industry has allowed designers to become overly reliant on existing materials and the knowledge of their behaviour. While this can allow for efficient and optimised design, it can prevent possible adoption of new innovative and regenerative materials.

10.3 VARIABILITY IN MATERIAL PROPERTIES

Materials are inherently variable, with a wide variety of properties that allow their use in different circumstances. These properties are often grouped together to help characterise and categorise materials. For the structural engineer, it is often the mechanical properties (compressive, flexural and shear strength, and stiffness) that are the primary concerns, followed by others that could include durability, thermal performance, fire and environmental impact. Within the design, it is typical to use only a single value for each property. However, there is always some variability in the definition of material properties. This intrinsic variation in measured performance is a key parameter in increasing confidence in the use of the material, and is considered within the conventional limit state design approach, as well as by improving and refining production techniques to manufacture a more consistent material. The variability in material properties is attributed to the physical and chemical make-up of the materials; so, understanding material science can lead to understanding material properties better.

10.3.1 The material science of material properties

The constituent materials and the manufacturing process will result in a material with specific physical and chemical properties. Modifying the variables can result in improved performance and is often the focus of material scientists, researchers and manufacturers, with structural engineers often working in collaboration with these disciplines. However, the starting point for a structural engineer is consideration of the uniformity of the material and how this categorisation can infer potential differences in properties. These aspects include homogeneity and the related isotropy. Homogeneity refers to the uniformity of the material structure while isotropy refers to the uniformity of the physical properties with respect to the different directions. While there is a clear definition between the two, the terms are often (and arguably incorrectly) used interchangeably by structural engineers. Often, the inference is that homogeneous materials result in isotropic properties, which, for many structural engineering materials, is broadly a suitable assumption.

- **Homogeneous materials**: A material with a uniform composition and structure throughout. When the material is a composite, typically it implies an equal distribution of the components throughout the material. Examples of homogeneous materials are typically metals, alloys, ceramics and resins
- **Heterogeneous materials**: A material with non-uniform composition and structure throughout. Examples include many natural building materials, such as timber and bamboo, but also include different varieties of fibre polymer composites

The added complexity to this simple categorisation is that some materials could be considered both heterogeneous and homogeneous, with the most common example being concrete. This often depends on the scale under consideration and how the definition is interpreted. Literally, concrete is a mixture of cement, water and a range of aggregate sizes; therefore, it is a heterogeneous material without a uniform composition. At the macro and meso scale, it could be considered homogeneous and at least continuous. However, observation would show clearly discrete and non-uniform particles at the smaller micro and nano scales. This appreciation of the potential varied nature of concrete should be applied to any other novel or non-conventional material. The multi-scale morphology results in unique material properties that allow potential for structural engineering applications. The broad nature of these properties is often first considered with respect to the isotropy; the variation in physical properties in different directions and as such materials can be described as:

- **Isotropic materials**: A material that has identical properties in all directions. Typical examples include glass and metals
- **Anisotropic materials**: A material where the physical properties will vary in different directions. Typical examples include timber, where engineered timber

often utilises the benefits of timber's natural anisotropy to result in tailored directional performance

- **Orthotropic materials**: A material where the properties differ along three mutually orthogonal planes (typically based on cartesian or polar axis). These are a subset of anisotropic materials, with timber typically considered to be orthotropic

Thinking again about concrete, it is often considered (especially for structural design purposes) to be isotropic, very rarely concerned by the member's orientation from a material perspective. Concrete's isotropic structural behaviour can be observed when concrete cubes are tested. The orientation of the cube is rarely specified or even considered during testing, as the results are often equivalent irrespective of the orientation. Again, this is associated with the context of the scale, and for engineering applications it is reasonable to assume the material is homogeneous. However, understanding the make-up of concrete at different scales can result in different properties, such as the transport properties of water, gases and ions, for example (Figure 10.2).

Figure 10.2: Existing concrete beam, sawn through and polished revealing the distribution of aggregate and binder

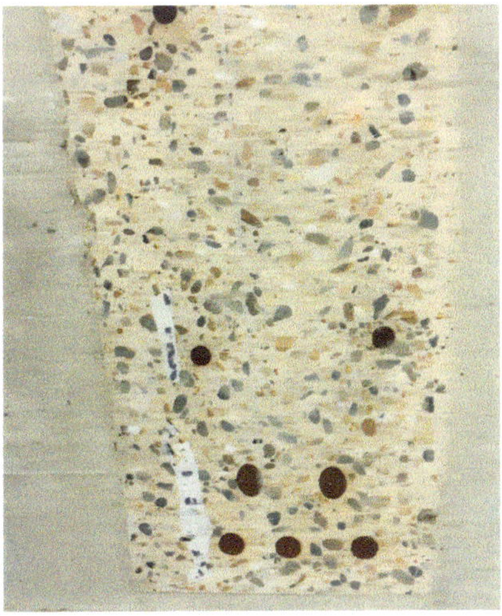

When we consider the design of structures, we must consider the material and its particular physical properties. As seen with concrete, one of the most well-used materials has greater intricacies owing to material science. For novel non-conventional materials, this variation in material properties can be more pronounced. Therefore, an awareness of the potential for variation in homogeneity and isotropy should be considered, as it will impact the performance, analysis, construction, design and testing.

10.3.2 Characterising materials' performance

Materials are typically tested to understand their behaviour or measure their performance (Chapter 11). The experimental material properties are more or less variable depending on the nature of the tested material or by the typology of conducted test. A probability distribution allows the modelling of the probability of a random variable with a simple parametric functional form. This section presents a few probability distributions, referring to the models commonly used in civil engineering applications.

10.3.2.1 Uniform distribution

When there is no information about a specific variable but only the two possible extreme values of the range, the ideal solution is to give an equal (uniform) probability to all values in the predefined range. This distribution is known as uniform distribution, and its two parameters are a and b (Figure 10.3), i.e., the maximum and minimum value, respectively. The probability density function (PDF) (Figure 10.3a) is constant, and the cumulative distribution function (CDF) (Figure 10.3b) is a linear function equal to 0 at a (and all the values lower than a) and equal to 1 at b (and all the values larger than b). A typical case of when this can be used is when we know only a specific variable's minimum and maximum values. More information is given on PDF and CDF in Appendix A.

Figure 10.3 (a) PDF and (b) CDF of a uniform distribution

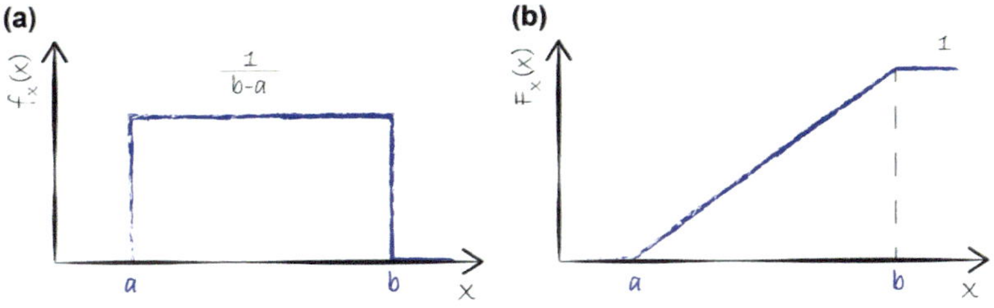

10.3.2.2 Normal (or Gaussian) distribution

While we may specify a strength for a given material, it is unlikely that we achieve that strength every time we test it. It is common that strength measurements will often cluster around the mean with fewer and fewer measurements further away from the mean. This leads to the frequently utilised normal distribution (Figure 10.4), which is characterised by the mean and standard deviation. These values are often utilised within conventional structural design to establish the characteristic strength, which is then further modified by the material safety factor.

Figure 10.4 (a) PDF and (b) CDF of a normal distribution

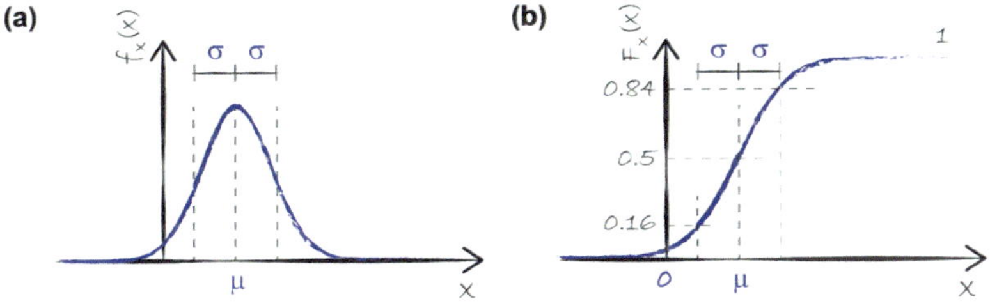

The normal distribution is the most well-known and most used distribution to characterise a physical parameter defined in the domain of real numbers. The PDF of the normal distribution is:

$$f_X(x) = \frac{1}{\sqrt{2\pi} \cdot \sigma} exp\left[-\frac{(x-\mu)^2}{2\sigma^2}\right] \qquad 10.1$$

where the parameters μ and σ are the mean and standard deviation, respectively. The distribution is symmetric with respect to the mean (i.e., μ), and the spread of the bell-shaped function is controlled by the standard deviation (i.e., σ). Figure 10.4a provides a graphical representation of the PDF and its parameters. Figure 10.4b shows the corresponding functional form of the CDF; it has a typical 'S' shape, and the μ – σ, μ and μ + σ correspond to the 16th, 50th and 84th percentile, respectively.

This distribution is widespread due to its practical applications and the central limit theorem, which demonstrates that the sum of independent random variables tends to a normal distribution even if the original distribution of the variables is not normal. With respect to the sum of normal random variables, such as the linear combination $Y = a_1X_1 + a_2X_2 + \ldots + a_nX_n$, the mean and the standard deviation of the sum (i.e., Y) can be calculated as follows:

$$\mu_Y = a_1\mu_{X_1} + a_2\mu_{X_2} + \ldots + a_n\mu_{X_n} \quad 10.2$$

$$\sigma_Y = \sqrt{\sum_{i=1}^{n} a_i^2\sigma_{X_i}^2 + 2\sum_{i=1}^{n}\sum_{j=1}^{n}\rho_{i,j}a_ia_j\sigma_i\sigma_j} \quad 10.3$$

where μ_{Xi} and σ^2_{Xi} are the mean and variance of the generic variable X_i, respectively. The term $\rho_{i,j}$ is the coefficient of correlation between the variable X_i and X_j if a correlation exists. In the second addendum of Equation 10.3, some subscripts have been removed for the sake of notation simplicity. The property presented in Equations 10.2 and 10.3 is very important for classical reliability analyses.

Referring to the case study in Chapter 9 at the end of Part 2, we can see that dimensions, loads and density have all been assumed to be represented by the normal distribution.

10.3.2.3 Standard normal distribution

The standard normal distribution (Figure 10.5a) is a special case of normal distribution obtained by fixing the mean equal to 0 and the standard deviation equal to 1. The functional form of the CDF for a standard normal distribution is usually indicated with $\Phi(\cdot)$. A random variable that can be described with a normal distribution can be identified as a normal random variable. The standard normal random variable is obtained as:

$$U = \frac{X - \mu}{\sigma} \qquad 10.4$$

The normal distribution can also be multi-dimensional or multivariate. A normal multivariate distribution (Figure 10.5b) can be used to model jointly distributed normal random variables. The PDF of such a distribution is:

$$f_X(x) = \frac{1}{\sqrt{(2\pi)^k |\Sigma|}} exp\left[-\frac{1}{2}(X-\mu)^T \Sigma^{-1} (X-\mu)\right] \qquad 10.5$$

where k is the number of variables considered (e.g., two in Figure 10.5b), $\boldsymbol{\mu}$ is the vector of the mean values for the considered values and Σ is the covariance matrix. The covariance matrix defines the correlation structure among the variables.

Figure 10.5: (a) Standard normal distribution (b) joint normal distributions with and without correlation

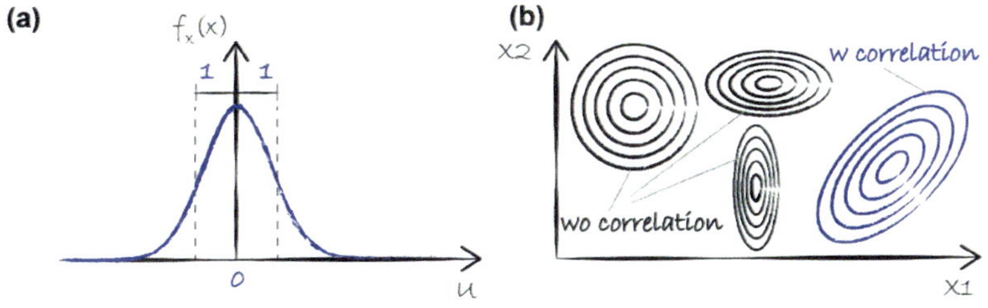

In Figure 10.5 the presence of correlation (blue lines) is clear when the variables relate to each other via a mathematical (linear or non-linear) relationship. The lack of correlation (black lines) is such that knowing one variable does not give information about the other.

The normal distribution is only representative when we consider the 'population'. A population in statistics encompasses a complete data set. For example, if we are considering concrete compressive strength testing, the population could be considered all the concrete used in the building with the same specification. As such, this makes the determination of the mean compressive strength practically impossible; we would need to test (and break) every bit of concrete to determine the population's average (and distribution) strength. This would result in no longer having a building. Common practice is to take samples of the concrete to test and determine the strength properties. A 'sample', in statistics, is a subset of the population. Intuitively, if only one concrete sample is tested, we may not have a lot of confidence in how representative that is of the whole structure (our population); therefore, we ideally want to take into account that we are only ever testing a sample (a subset of the whole) and the normal distribution may not be our best approach.

10.3.2.4 Student's t-distribution

The normal distribution assumes that the entire population is represented, which is intrinsically not the case when we establish the values based on observed data (typically from material testing). Whenever a material property is established empirically it is always based on a randomly-selected subset of a population. This results in an inherent uncertainty in the results derived, based on the representation of the population sample. For example, structural design starts with calculating the mean value, a singular value established from empirical testing. However, if we were to test more specimens (even just one more), then intrinsically, the new value obtained would change our originally calculated mean. The more we test, the closer the calculated unknown parameter (our mean) of the subset will approach the population parameter (our population mean), because our subset becomes ever more representative of the population.

The t-distribution (often referred to as Student's t-distribution) (Figure 10.6) is similar to the normal distribution, being symmetric and bell-shaped, but the exact shape depends on sample size. With a smaller sample size, the bell shape is spread out further, implying a greater distribution spread. As the number of specimens increases the resulting distribution tends towards the normal distribution. As with the normal distribution, confidence intervals can be calculated but will have a greater range.

Figure 10.6: Student's t-distribution. ν is the degree of freedom

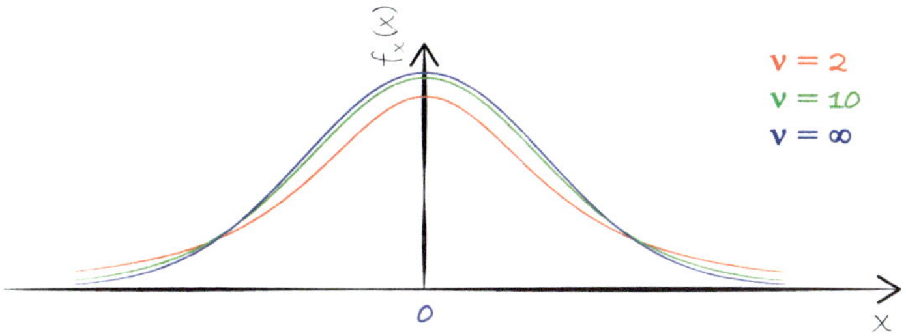

The t-statistic is similar to the standard normal variable in Equation 10.4, with the difference of the denominator being the standard deviation (s) of the sample rather than the population (σ):

$$t = \frac{X - \mu}{s} \qquad 10.6$$

The PDF of the t-distribution is:

$$f_X(t) = \frac{\Gamma\left(\frac{\nu+1}{2}\right)}{\sqrt{\pi\nu}\,\Gamma\left(\frac{\nu}{2}\right)}\left(1 + \frac{t^2}{\nu}\right)^{-(\nu+1)/2} \qquad 10.7$$

where ν is the number of degrees of freedom and Γ is the gamma function.

10.3.2.5 Confidence intervals

This, therefore, calls into question how representative the sample mean is of the population mean. From a safety perspective, the sample mean, as calculated from testing, and used for the basis of design would very likely differ from the population mean. Therefore, it is often prudent to consider that the actual population mean could exist within a range of possible values, known as an interval estimate. Commonly, confidence intervals (Figure 10.7) can be used with the observed data from testing, with the interval associated with a confidence level (a somewhat subjective value). It is common to take a 95% confidence interval for many structural properties, meaning that there is a 95% probability that the calculated confidence interval from future testing encompasses the true value of the population parameter. Therefore, the confidence interval is considered as a measure of the accuracy of the estimation.

Figure 10.7: Samples (of different size) from the same population with confidence interval (CI)

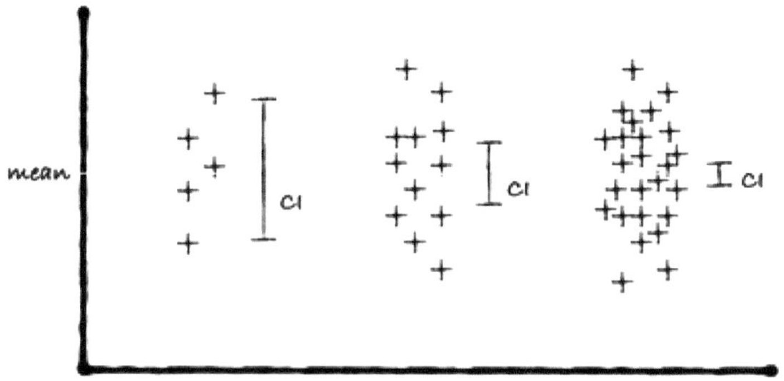

As seen by considering confidence intervals, the amount of confidence we have in the measured material properties is dependent on the amount that we are able to test. We are unable ever to truly test enough to establish the parameters of the population and establish the normal distribution. Typically, when establishing material properties, only a limited number of specimens can be tested, often due to cost, time and other practical constraints. When the sample size is small (typically fewer than 30), and the population's standard deviation is unknown (as from physical testing), then the Student's t-distribution can be used.

10.3.2.6 Lognormal distribution

A clear shortcoming of the normal distribution and t-distribution is that the curve is always symmetric around the average, and as such implies that there exists the same probability of occurrence for the performance equally above and below the average. By extension, these distributions approach infinity (in the positive and negative direction) so there will always be some probability of negative values. For some characteristics, this is simply not possible, while for others, it becomes non-sensical. As an example, density can be established by measuring both mass and volume and can vary between specimens of the same material, but it will always be positive; it certainly isn't possible to have a negative mass or a negative volume both logically and mathematically. However, from a statistical view, if a material has a low density with significant variability, then it can result in a characteristic value of a negative density. Many material properties within structural engineering also imply direction or at least relative direction. Depending on the direction of uniaxial force with a material, tension or compression can result, and while different disciplines will infer different sign conventions, the opposite intrinsically infers the opposite force. Therefore, when a material is tested in, say, tension (where tension is assumed positive), again statistically, we could end up with a negative result. This would not represent a compression strength purely based on the sign.

The lognormal distribution is suitable for random variables whose logarithm is normally distributed. This distribution is often used for variables that can be either only positive or only negative. The PDF of the lognormal distribution is:

$$f_X(x) = \frac{1}{x \cdot \sqrt{2\pi} \cdot \xi} exp\left[-\frac{(\ln x - \ln \eta)^2}{2\xi^2} \right] \quad 10.8$$

where ln is the natural logarithm, the parameters η and ξ are the median and the logarithmic standard deviation, respectively. The distribution is asymmetric, presenting a right skewness (Figure 10.8a). Figure 10.8b shows the CDF; the three most important percentiles are also illustrated.

Figure 10.8: (a) PDF and (b) CDF of a lognormal distribution

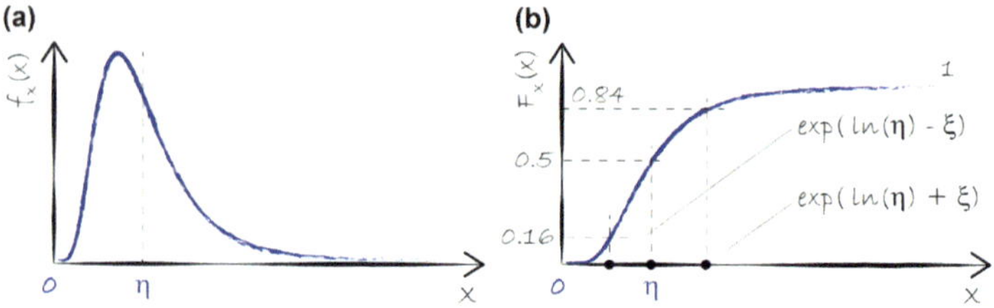

Often, in the literature, η and ξ are not provided, and the μ and σ of the associated normal distribution are provided instead. In such a case, the parameters of the lognormal are:

$$\xi = \sqrt{\ln\left[1 + COV^2\right]} \quad 10.9$$

$$\eta = exp(\ln\mu - 0.5 \cdot \xi^2) \quad 10.10$$

where *COV* is the coefficient of variation equal to the ratio σ/μ. This distribution is very common since the central limit theorem still stands for the product of variables whose logarithm is a sum. Also, this distribution is very useful for random variables that always assume only one sign (either positive or negative), which is the case in many physical/engineering applications (e.g., compressive strength, tensile strength). For example in the case study at the end of Part 2 the strength, shear and elastic modulus are all modelled as lognormal distributions (Table 9.2).

10.3.2.7 Generalised extreme value distribution

The concept of structural safety is based on the maximum load experienced over the structure life cycle not exceeding the minimum resistance. Probabilistic distributions capable of modelling such extreme values can be very useful for structural reliability assessments. The generalised extreme value distribution models the smallest or largest value among a large set of independent, identically-distributed random variables. This distribution is three-parametric and depends on the location parameter θ_1, the scale parameter θ_2 and the shape parameter θ_3. This model brings together in a simple analytical formulation three different models (Figure 10.9): Gumbel ($\theta_3 = 0$), Fréchet ($\theta_3 > 0$) and Weibull ($\theta_3 < 0$), also known as type I, II and III extreme value distributions. The generalised extreme value distribution lets the data decide which distribution of the three is suitable. The PDF of the function is:

$$f_X(x) = \frac{1}{\theta_2} \times exp\left[-\left(1+\theta_3 \times \frac{x-\theta_1}{\theta_2}\right)^{-\frac{1}{\theta_3}}\right] \times \left(1+\theta_3 \times \frac{x-\theta_1}{\theta_2}\right)^{-1-\frac{1}{\theta_3}}, \text{for}\, \theta_3 \neq 0 \qquad 10.11$$

$$f_X(x) = \frac{1}{\theta_2} \cdot exp\left[-exp\left(-\frac{x-\theta_1}{\theta_2}\right) - \frac{x-\theta_1}{\theta_2}\right], \qquad 10.12$$

$$\text{for}\, \theta_3 = 0$$

The weakest link theory and Weibull distribution are typically utilised within many fibre polymer composites and fracture mechanics, such as designing with glass or stone[10.4].

Figure 10.9: The three functions belonging to the generalised extreme value distribution

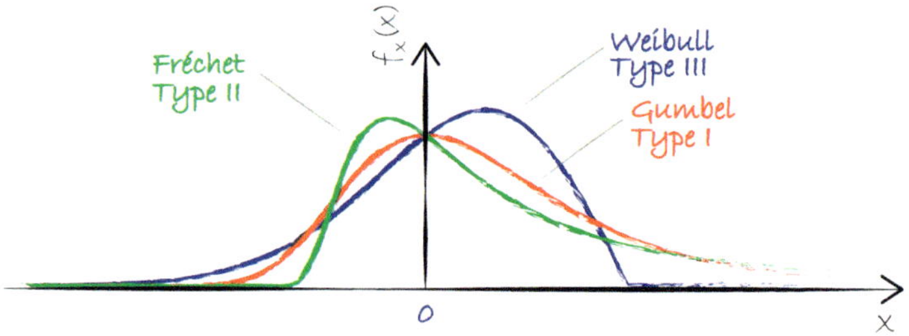

This distribution family is useful in engineering for describing extreme events of particular interest for designing safer infrastructures (e.g., the bending moment induced by vehicular traffic on bridge decks[10.5]).

10.3.3 Material equivalence and difference

For early development and use of new materials it would be sensible to consider how an unknown material compares to something more familiar or a given value. For example, how do two strengths compare? While this should consider the full material behaviour, comparing the quantitative material properties can be more difficult given the inherent variability observed through experimental testing. When using a single parameter of a material for the design, such as strength, an intrinsic assumption of this approach is that the material property effectively represents all the material being used and therefore the population. Comparing materials could be as trivial as determining the relative value of a given property. As an example, typically for structural application the material with the highest strength is most desirable. However, when considering a new material that has limited test data and the knowledge that the tested sample doesn't represent all of the material (the population), then it is too simplistic to compare single values. It is often worth considering whether two materials can be described as statistically different, or statistically equivalent.

The t-distribution is the basis of many statistical hypothesis testing analyses which can be used in evaluating how the properties of two materials compare; accounting for the sample size. Using the Student's t-test allows for comparison of two sample means which can demonstrate statistical significance, difference or equivalence, allowing for material property comparison.

The t-test makes the same assumptions about the data in that they:

- Are independent
- Are (approximately) normally distributed
- Have a similar amount of variance within each group being compared

If this is not the case then an alternative would need to be used, such as the Wilcoxon Signed-Rank test for data with unequal variances[10.6].

Along with the confidence interval, a one-sample t-test allows for the comparison of the sample mean compared to a specified value.

There are a few types of available t-tests which will depend on whether the groups being compared come from a single population or two different populations, and whether you want to test the difference in a specific direction.

The first will determine if we are considering a paired t-test, two-sample or one-sample t-test.

- **A paired t-test** is when the groups come from a single population and can often determine whether an intervention has any impact. As it requires the literal same specimens to be used, it is relatively uncommon in structural testing, due to its generally destructive nature

- **A two-sample t-test** (also referred to as an independent t-test) is when the groups come from two different populations; such as two different materials or mix designs, etc.
- **A one-sample t-test** is used when comparing one group against a single value. This value could be given from a relevant standard or be a minimum or maximum value. For example, by ensuring a minimum strength is achieved or having confidence in not exceeding a specified embodied carbon value

The second choice determines whether you are concerned with showing that there is just a difference, or if you are interested in knowing the relative nature of the two samples. Referred to as two-tail or one-tail t-testing, the terms come from considering both or just one of the tails of a distribution.

- **A two-tailed** t-test is used when you are concerned by whether the two populations are statistically different from each other. For many structural applications this is particularly useful, and the mean values determine the relative nature
- **A one-tailed** t-test is used when you only want to know whether one population mean is greater than or less than the other. This is particularly useful when coupled with a one-sample t-test, to ensure that the sample is greater than the minimum value or less than the given maximum value

The t-test estimates the difference between two group means based on the ratio of the difference in the group means, against the pooled standard error of both groups. To quantitively assess this, it is set up as a form of hypothesis testing to determine the statistical significance that gives a 'p-value' based on the following hypothesis:

- The null hypothesis (H_0) is that the true difference between these group mean values is 0

- The alternate hypothesis (H_a) is that the true difference is different from 0

The resulting p-value is compared to an α value, typically 0.05 and if it is smaller we can **reject the null hypothesis** of no difference and say with a high degree of confidence that the **true difference in mean values is not equal to 0**. Therefore, there is a statistically significant difference. The 0.05 relates to a threshold of a 5% chance that the null hypothesis is actually correct. While we can set any value to be this threshold, convention uses this 0.05 α value. It is worth noting that if this test results in a p-value greater than 0.05, while it is appropriate to declare that the two samples are not statistically different, it is not true that they are statistically equivalent. This is based on the initial null and alternative hypotheses not being appropriate to establish if the mean values are equivalent.

A variation of hypothesis testing would be to undertake specific equivalence testing. While there is likely to be some variation from two samples, this variation must be within acceptable bounds. A common engineering example is in the casting of concrete samples. Imagine six samples were cast from one larger concrete mix and there was no difference in the conditions that would lead to a variation in strength development. While theoretically they should all have the same compressive strength, intrinsically, from engineering experience we know that there will be some variation, and they won't 'literally' be the same. If we divided the original six specimens into two groups we could then calculate the mean, which would likely be slightly different. From this real-world example, we would clearly expect these two groups to be equivalent even though numerically the means are different. Defining the bounds is to define how much the 'slight difference' can be before we would say they are not statistically equivalent. Similarly, defining something as not statistically equivalent does not define them as statistically different.

Similar to the testing of difference, to assess quantitively for equivalence we determine the statistical significance that gives a p-value. This p-value is compared again to the threshold value, again typically taken as 0.05. This testing is based on the following hypothesis:

- The null hypothesis (H_0) is that the true difference between these group means is outside the equivalence interval
- The alternate hypothesis (H_1) is that the true difference is inside the equivalence interval

A common method of undertaking the test is the 'two one-sided t-tests' (TOST) procedure. This results in two separate hypothesis tests based on the one-sided t-tests, with the null hypothesis needed to be rejected for both:

- H_{10}: The difference is less than or equal to the lower limit for equivalence
- H_{1a}: The difference is greater than the lower limit for equivalence

- H_{20}: The difference is greater than or equal to the upper limit for equivalence
- H_{2a}: The difference is less than the upper limit for equivalence

While we can define what the equivalence limit is, typically it is often based on confidence intervals of $1–2\alpha$. This is because we are considering two confidence intervals to determine equivalence. Given that typically α is taken as 0.05, this results in a 90% confidence interval.

These statistical methods allow for greater confidence in how a new material may compare against more established materials or specific requirements. These methods require establishing the mean and variance of two samples, with one of the samples potentially being a conventional material.

While a conventional material often has a significant amount of data available, it may still be worth testing. Many methods of testing will have systematic errors, which the design standards attempt to reduce through consistent methods, but there may be some variation between different test sites that would hinder the statistical analysis.

10.3.4 Engineering equivalence

While comparing the means can demonstrate a novel material is equivalent statistically, a typical engineering design approach could result in different properties being used. As discussed in Part 2, rarely is the mean value used for design. Typically it is converted to a characteristic value by considering the standard deviation.

Interrelated with the natural variability of the materials is the uncertainty of the material, which is accounted for in design by the partial material factor. While this can be determined from a reliability or probabilistic analysis, for conventional materials it is given in codes. These partial factors are typically determined based on expert judgement, historical developments from previous codes and calibrations with experimental work. Novel materials clearly lack this historical context, and experimental work may be limited. As a result, partial factors cannot be applied and a reliability analysis should be carried out.

However, at the conceptual design phase, where decisions need to be made quickly, using partial safety factors for similar materials can help with material selection and initial sizing, but caution needs to be applied.

As an example, rammed earth has a material factor which is often suggested to be between 3 and 6[10.7], dependent on the contractor's experience and how compliant the material is to the recommendations. At an early stage of design, it would be logical to use the higher value that allows for a conservative design. While this logic cannot be entirely criticised, it can have a significant impact on the design of a wall, where the wall thickness would have to double for the higher material safety factor for the same load carrying capacity. This increase in wall thickness could then deem the design, and ultimately the material, unsuitable before refinement can begin. However, the converse could result in the design not being feasible due to cost/time/ability to find a suitable contractor to meet the resulting design specification that underpins the higher material performance.

While it may be tempting to compare materials based on the design property, this results in an artificially imposed optimisation criteria. As an example, considering only the design structural strength would imply that the material with the highest value would be preferred. However, generally, key to structural design is load-bearing resistance and it is therefore a factor of both strength and geometry of the member (e.g., cross-sectional area assuming just compression/tension). Therefore, a higher strength would allow for a smaller cross-section and therefore greater material efficiency. Using less material clearly feels advantageous because the materials are fundamentally different. Other parameters, for example, environmental impact, cost and other engineering properties, may mean that an ideal solution is one with a lower strength and a bigger cross-section. Therefore, having a broader understanding of other material properties and the resulting impact on the structure should be considered.

10.4 PROPERTIES

10.4.1 Assessing materials

As we have seen, there are lots of materials that could be suitable for building applications, and each has lots of disparate properties that can vary in different ways. While it is trivial to compare one material's property to another to establish which is 'stronger' or 'stiffer', material properties can't be considered in isolation. This is intrinsic to structural engineering when we use limit state design, typically designing for strength and stiffness differently. However, beyond these mechanical properties that are often the focus of structural engineering, multiple properties should be considered and can become critical at different stages.

Note: The Grenfell Inquiry Report recommendations list understanding materials as one of the four key safety criteria for delivering safe buildings[10.8].

The significance of the properties will depend on the material's behaviour under a variety of conditions. Codes and standards may prescribe different properties to be investigated or classified and are often expressed within a 'declaration of performance'. These may indicate both the properties and relevant test methods that typically need to be considered, especially for the development of new materials that are looking to be certified eventually. While compliance to many standards and guides may give confidence to the designer and are often recommended, they are frequently not an absolute requirement for a country's building regulation. Any prescribed list is unlikely to consider the full array of properties that could or should be investigated. Many of the prescribed property requirements are based on assumptions over the materials that are likely to be considered, typically similar to conventional materials. This can be a significant shortcoming where the behaviour of a material is fundamentally dependent on a range of parameters that is not considered for other materials. For example, timber's behaviour is known to vary significantly based on the moisture content and therefore the environment. In this case the codes and standards allow for this. However, if an unconventional material behaves similarly to moisture (or other parameters) this may never be considered or investigated. For example, rammed earth is often compared to a concrete wall, due to similarity in form and construction. However, not only is the strength considerably different, how the strength relates to moisture content is also more significant.

Recognising the various dependencies is often a major challenge and can lead to a greater risk (real or perceived) of new materials being underutilised. One approach is to consider the Johari window (Figure 10.10). Developed from a basis in psychology, it can be a useful tool to consider in engineering when considering one's own understanding of materials. Properties of a given material can be positioned in one of the quadrants based on understanding of the specific material and potential aspects from a broader understanding of materials. However, there will always be the 'unknown unknown' quadrant that remains a concern with field and *in situ* testing, as discussed in Chapter 11.

Figure 10.10: Johari window for material assessment

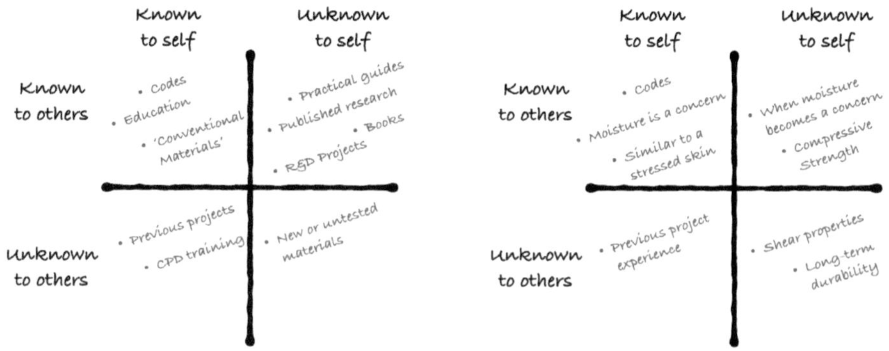

a) General assessment for construction materials **b)** Specific assessment example for straw-bale

10.4.2 Mechanical properties

The foundation of structural analysis is fundamentally based on material behaviour, specifically the relationship between stress and strains, referred to as the 'constitutive laws'. The generalisation of the material's stress-strain relationship into elastic/perfectly-plastic behaviour leads to the basis of both elastic and plastic analysis shown in Figure 10.11.

Figure 10.11: Elastic/perfectly-plastic behaviour relationship

Elastic analysis assumes the relationship for a given material is linear and reversible before yielding or cracking. The generalised three-dimensional relationship as given by Hooke is:

$$\sigma_{ij} = C_{ijkl}\varepsilon_{kl} \quad 10.13$$

where C_{ijkl} is the fourth-order elastic tensor that is dependent on material properties. Indices i, j, k, l can be 1, 2, 3 and represent three orthogonal spatial axes, x, y and z, resulting in the elastic tensor having 81 components in three dimensions. This represents a potential 81 theoretically different material properties just for this singular relationship. Typically, these 81 components can be reduced depending on the isotropy of the material. These are often reduced to 21 components for anisotropic materials, nine components for orthotropic materials and down to two when isotropic. This leads to the various relationships between elastic moduli for homogeneous isotropic materials, using bulk modulus, Young's modulus, Lamé's first parameter, shear modulus, Poisson's ratio and P-wave modulus, where only two parameters are needed to establish the others.

While the elastic/perfectly-plastic behaviour is typically assumed, the stress-strain behaviour of real materials can be quite different. Even for conventional homogeneous isotropic steel the actual stress-strain behaviour demonstrates the additional complexities of actual materials and how they deviate from the generalised assumption. Steel shows strain hardening, whereas concrete exhibits strain softening and also leads to the classification of brittle and ductile materials (Figure 10.12). Even this stress-strain behaviour is not a 'true' behaviour but what is commonly referred to as an engineering stress-strain curve. The engineering stress-strain curve assumes the cross-section of materials stays constant, however it rarely does. This is observed through the 'necking' process of steel in tension and the crushing of a concrete cube that can lead to a cone-like shape. Concrete demonstrates the potential for materials with a non-linear stress-strain behaviour still to be utilised (Figure 10.13).

Figure 10.12: Engineering stress-strain curve for steel

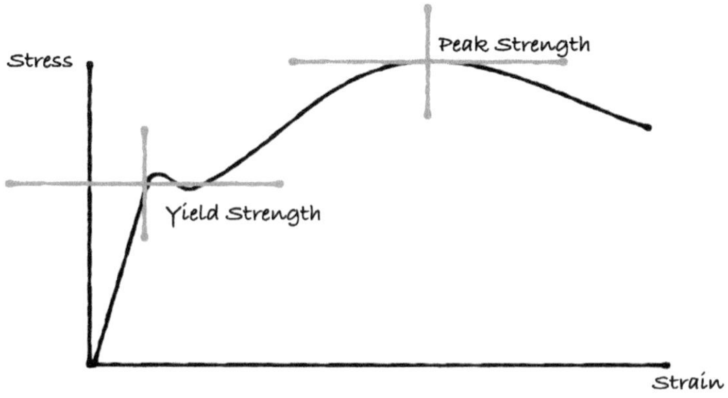

Figure 10.13: Engineering stress-strain curve for concrete

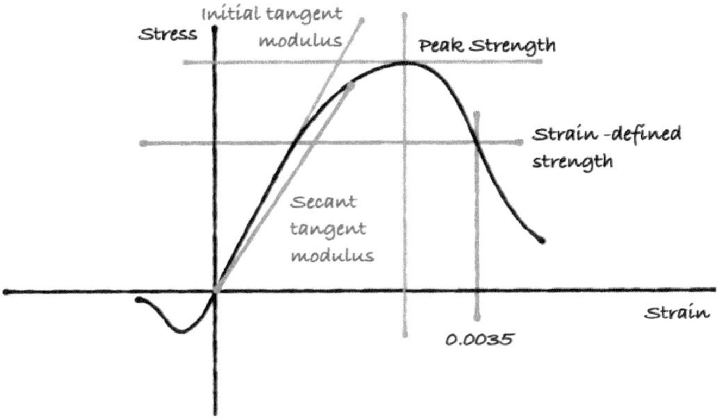

Structural design is often concerned by stress-strain behaviour under tension, compression and shear and will satisfy the requirements of the stress tensor. However, it is also common to investigate the generalised engineering condition of bending resulting in a moment and rotation. The shape of the curve demonstrates the overall behaviour of the material under specific loading. However for structural design, this behaviour needs to be distilled down to quantifiable properties. Typically, this will be a stress value that represents the strength of the material and the elastic modulus.

The elastic modulus is defined as the slope of the stress-strain curve. While this is linear for the idealised material and elastic material such as steel, many materials do not exhibit a clearly linear elastic modulus, e.g. concrete. As concrete demonstrates, we can still approximate the elastic modulus through calculating the gradient of the stress-strain behaviour. Typically, for both steel and concrete, the initial linear portion of the curve is used with the steepest slope calculated. However, depending on the material, this may not always be suitable and other methods of calculating the elastic modulus may be more appropriate (Figures 10.13 and 10.14):

- **Initial tangent modulus**: the slope is calculated on the initial linear portion of the stress-strain curve. Typically the steepest slope is reported as the modulus
- **Secant modulus**: the gradient between the zero stress-strain point and another point (e.g., 90% maximum) on the curve
- **Chord modulus**: the modulus is calculated as the gradient between two strain points
- **Segment modulus**: the modulus is calculated based on a gradient between two given points. Unlike the chord modulus a gradient is calculated based on least squares regression at any point between those selected
- **Tangent modulus**: a slope is calculated as the tangent at any point on the stress-strain curve

Figure 10.14: Engineering stress-strain curve for straw bale

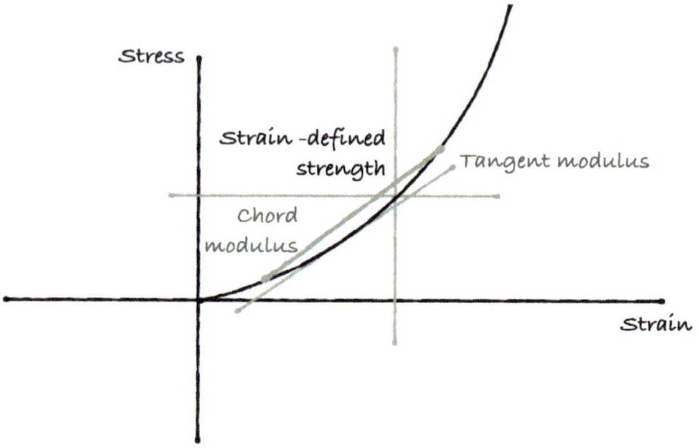

It may seem that the strength of material is trivial to measure, especially under an ultimate limit state. Often it is taken as the maximum stress value obtained during testing. However, this is not always suitable for many materials where a lower than peak stress is used as the strength, or a clear peak is not obtained. This approach is already commonly used. As an example, the peak stress for steel is not taken, instead taking the yield point. The choice to take a non-peak stress value is typically dependent on the stress-strain curve behaviour obtained through observation of the specimen. There are several possibilities (Figures 10.12–10.14):

- **Peak strength**: the maximum stress recorded and taken as the strength of the material. The strength of concrete is based on the peak strength recorded
- **Yield strength/elastic limit**: the stress a material can withstand without permanent deformation. The strength of steel is calculated based on its yield strength
- **Proportional limit strength**: the strength at which the amount of stress stops being linear proportional to strain. This may be based on assumptions of how the elastic modulus is calculated. While often similar to the elastic limit, they can be different. Many elastomers, such as rubber, show non-linearity up to the elastic limit where permanent deformations would occur
- **Strain-defined strength**: the stress for a given strain value. This is often used to ensure excessive deformations do not occur. A common example is that concrete's defined strength is often limited. If it hasn't reached peak before the strain value, this strain value will vary based on national codes and standards but is typically in the range of 0.002 to 0.0035. Insulation materials often use a strain value of 0.1 for a compressive strength, but these are for non-loadbearing applications. The resulting strength is clearly dependent on the stiffness of the material and judgement over the acceptable strain and allowable deformation. While this is linked to serviceability requirements, the strength should still be treated as an ultimate limit for the purpose of design. The limit on strain for structural applications is often taken as 0.01
- **Observation-defined strength**: the strength based on an observation from testing which may be visible on the stress-strain curve. An example would be the observation

of initial cracking, which may not significantly impact the stress-strain behaviour but is likely to be undesirable and should be limited. This approach is clearly dependent on a level of expertise and interpretation, and may be less repeatable and result in greater variability when multiple specimens are investigated

The stress-strain behaviour is key to understanding the behaviour of a material and is necessary for its specification within structural applications. This behaviour may vary under tension, compression, shear and bending and should be investigated independently to help understand the structural potential of a material. In addition to this behaviour there are numerous other mechanical properties that it may be prudent to investigate. This may include, but is not limited to: hardness, toughness, impact strength, fracture toughness, fatigue and creep.

- Hardness is the resistance of a material to localised and permanent deformation. Hardness can refer to deformation caused from indentation or abrasion including scratching and cutting. Hardness is dependent on the stress-strain behaviour and other characteristics. There are different categories of measurements of hardness that include scratch hardness, indentation hardness and rebound hardness with a variety of test methods for each
- Intrinsically related to the stress-strain curve is the toughness of the material. Toughness is the ability of a material to absorb energy and plastically deform without fracturing, and is commonly considered for metals. Toughness is the combination of strength and ductility and is calculated as the area under the stress-strain curve. A material is considered to be tough if it has both high strength and high ductility. Toughness indicates how much energy a material can absorb before

rupturing and is expressed as energy of mechanical deformation per unit volume prior to fracture. Toughness can be highly dependent on the rate of loading, temperature and the presence of any flaws. There are different approaches (including standard testing methods) to consider these variables

- The impact toughness, often considered as impact strength, is specially considered with a high loading rate. It is defined as the capability of the material to withstand a suddenly applied load and is expressed in terms of energy
- Fracture toughness is a material property that describes the material's capacity to resist fracture from a flaw within the material. Flaws are considered as cracks, voids, metallurgical inclusions, weld defects, design discontinuities or a combination. The fracture toughness of a material is defined as the critical stress intensity factor for immediate crack growth under given conditions
- Creep is a time-dependent deformation of a material while under a persistent stress that is typically below its elastic limit. It is often described as a 'viscoelastic' behaviour. The rate of deformation is intrinsically linked with the material properties but is also much more dependent on the temperature and moisture content (when the material is porous). A greater amount of creep is typically observed when temperature and moisture content are elevated. Creep can ultimately lead to failure, even though the stress conditions are below the maximum as derived from the stress-strain behaviour from uniaxial loading. In Figure 10.15, it is represented as strain as a function of time and there are often three stages identified — initially strain can occur relatively rapidly but will gradually decreases until it becomes approximately constant during the second stage but will then increase until failure

Figure 10.15: Creep behaviour

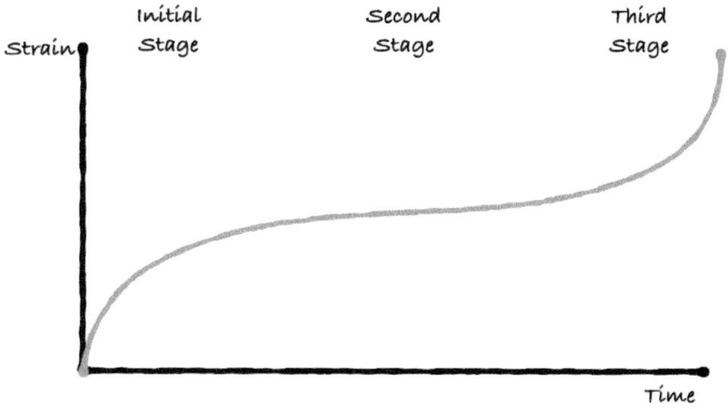

The impact of creep is common when designing with timber and accounted for in Eurocode 5[10.9]. With design of a timber structure, the additional deformation associated with creep is considered through modifying the stiffness by a k_{def} factor which itself is dependent on the actual material, duration of loading and service class. The service class is dependent on the environment and is a proxy of the moisture conditions. The impact of moisture and load duration on the performance of timber is not only seen in its creep behaviour but also its actual strength. The strength of timber is modified by a k_{mod} factor, again related to the service classes and load-duration classes. Many natural materials (and more specifically, porous building materials) show a similar variation in mechanical properties based on temperature and moisture content, so the potential impact should be considered and may need to be experimentally tested.

10.4.3 Hygrothermal properties

Hygrothermal properties are concerned with factors that influence the transport of heat and moisture through buildings. While typically a structural engineer may not focus on these properties, as we have seen, the temperature and significantly the moisture content can have a big impact on mechanical properties. The broader holistic understanding of building materials allows for overall greater material use.

Hygrothermal properties have significant impacts on wetting/drying and freeze/thaw cycles that can result in condensation, damp, mould growth and reduction in thermal resistance.

Thermal transmittance is a key parameter in the operational performance of buildings and therefore overall energy efficiency. Typically, thermal conductivity, which is a measure of the material's ability to conduct heat, is often used and indicates its effectiveness as an insulator. Typically, it is utilised when calculating overall thermal transmission, U-value (or the reciprocal, the thermal resistance; R-value), expressed over the surface area of a wall. While this is suitable for some construction systems it is not appropriate for systems based on frames. Alternately, Psi values are a measure of linear thermal heat transmittance that can often better represent a structural system and the potential thermal bridging.

While the thermal conductivity of a material is often required to meet many regulations, conventional heat transfer calculations are steady state. However, the temperature differential is often fluctuating. The thermal mass of a building material is often of interest as a passive method of storing heat resulting in dampening the temperature fluctuations. This can be beneficial to both the energy efficiency and the indoor thermal conductivity. Specific heat capacity is often established, which is used to calculate the impact of thermal mass.

While water can impact metal, commonly causing rust with steel, it can be detrimental in different ways to porous building materials due to its transport through the pore network. When considering water transport, it is often worth considering both liquid water and water vapour. The liquid water transfer properties include total water absorption and capillarity water absorption, which consider how much water can be stored as well as the rate of the absorption. Understanding how sensitive the materials are to these properties results in impacts to the overall design. Examples of the implications of these properties is the common practice of using a damp-proof membrane or designing timber connections to ensure that the timber is lifted off the ground. The transport of water vapour is akin to the transfer of heat. The water vapour diffusion resistance factor quantifies the steady state resistance of water vapour through the material. It relates to the property often referred to as 'breathability' which can be desirable for moisture-sensitive materials. This may seem counterintuitive to allow water vapour to move through the structure, but it will also allow any trapped moisture to move out. Porous material is also able to store water vapour within the porous network. Referred to as sorption isotherm or just an 'isotherm' (Figure 10.16) it indicates the equilibrium moisture content of the material at a given relative humidity of the surrounding environment. The isotherm behaviour can therefore indicate how mechanical properties may change with respect to the specific environment due to influence over the moisture content.

Figure 10.16: Sorption isotherm

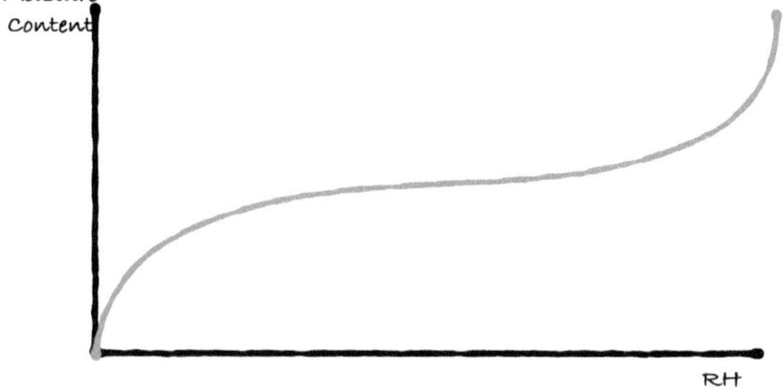

10.4.4 Material performance in fire

Following the Grenfell Tower tragedy engineers are again reminded of the importance of understanding how materials behave when exposed to fire. There are two distinct considerations.

Firstly, how combustible the materials are. In other words, if there is a fire will they burn? If they are combustible, we need to understand the rate at which they burn, how the material behaves during different phases of burning and how this impacts on the design. These are typically concerns for biomaterials, such as timber, bamboo and straw, but should be a consideration for any material.

Secondly, how does the material perform at elevated temperatures? Steel would need to be heated to a very high temperature to melt, but long before this the increase in temperature leads to a reduction in stiffness. Of course, some materials are insulating and/or act as a heat sink, so can provide a certain degree of protection.

Whenever looking at a novel material the expertise of a specialist fire engineer should be sought to ensure the design has fully considered the breadth of issues related to fire.

10.4.5 Environmental impact

Beyond thinking about material utilisation within an overall structural design, the environmental impact of the materials should be considered as part of a life cycle analysis (LCA). LCA is an analytical tool; it can also be used as a descriptive tool for the analysis of a consequences of choices. An LCA will typically consider several stages: product, construction process, use, end of life, and is further broken down into modules as in Figure 10.17[10.10]. Considering a material will typically involve modules A1–A3 and is often referred to as a cradle-to-gate analysis, but the choice of the material will often determine construction processes, in-use and operational impact and end of life.

While there is significant focus on embodied carbon, there are numerous ways that a material could impact the environment and typically these are combined into impact categories. An increasingly common approach when assessing materials is to make use of environmental product declarations (EPDs) which are based on product category rules. These rules dictate the categories and include:

- Global warming potential (GWP)
- Ozone depletion potential (ODP)
- Acidification potential (AP)
- Eutrophication potential (EP)
- Photochemical ozone creation potential (POCP)
- Abiotic depletion potential — elements (ADP)
- Abiotic depletion potential — fossil fuels (ADP)

Figure 10.17: Life cycle stages and modules

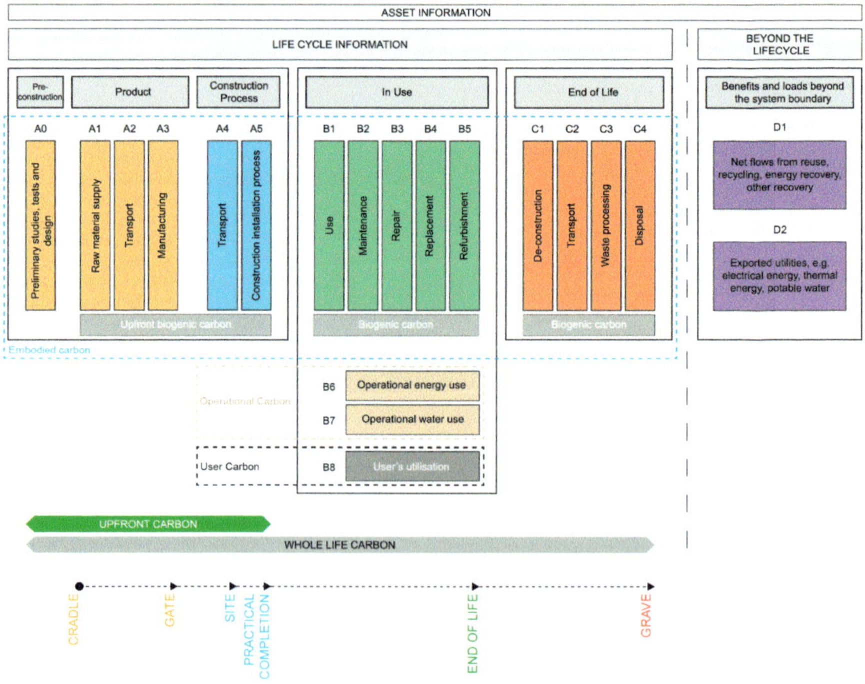

The declared unit for the impact may be given based on the volume or mass (or some other physical or geometric measurement). However, this is not always convenient for structural materials when the amount of material used is often designed based on the action resistance and strength of the material. Therefore, it would arguably be better to compare the impacts based on the action resistance or strength of the material. This defines the functional unit allowing for material comparison and better material selection. As for other properties discussed, the environmental impacts are often reduced to an average value that can be misleading when considering the variability. Beyond variability, a life cycle approach is also a good exemplar of uncertainty. As with traditional engineering design, assumptions are often made with life cycle analysis and these uncertainties also need to be accounted for. Beyond the material or individual product or system, it is increasingly common to consider the environmental impact based on floor area for a whole building which can allow for initial assessment during scheme design.

While the list in Figure 10.17 seems to be comprehensive, there are an increasing number of new parameters that reach beyond the product category rules and current LCA approach. Other areas such as biodiversity, human health and community can be further measured and evaluated to enable a regenerative approach to construction beyond one focused only on sustainability.

10.5 MATERIAL SELECTION

Material selection is one of several factors that can help influence sustainability in construction and is arguably one of the most important factors influencing embodied carbon in a building. Selection of building materials is common to all construction projects, but the decisions that guide these choices will vary in each situation. While building materials are required to conform to certain legislative requirements, for example, in the UK these include building regulations, energy-related product regulations and hazardous waste, there is still a wide choice of available products. Material selection is informed by multiple criteria, with a trade-off commonly occurring between performance, cost and environmental impact. The large number of materials and manufacturing processes available to engineers, coupled with complex relationships between the different selection parameters often make the selection process a difficult task. If the selection process is carried out wrongly, there will be a risk of neglecting a possible best material for certain applications. This risk can be reduced by adopting a systematic selection procedure. Taken a step further, the material selection (based on what is abundant) can inform the design brief.

Note: A number of case studies in *The regenerative structural engineer* highlight how material choice is a fundamental part of the brief, for example Yes Make's 'Using felled urban street trees, UK'. Case study 19[10.11].

Exploiting the full potential of materials is therefore key in material selection, particularly where they can provide multiple solutions. Currently material selection remains limited to specific performance criteria, and a narrow range of properties dominate the selection process. These performance criteria are those specified by legislative drivers as well as the client and design team, ensuring compliance, with the presence of additional non-mandatory performance criteria varying. This range of both mandatory and additional performance criteria inherently has varying units, drivers or performance references that make comparison of materials challenging. Material selection tools can allow for greater innovation, overcoming the common practice of selecting materials based on past experience alone.

10.5.1 Multi-criteria decision-making

There is a range of performance criteria to consider when selecting a material. Even when only considering legislature-compliant building materials, optimal selection will depend on maximising multiple desirable criteria, which can often be conflicting. Currently, selection by a designer usually relies on a judgement choice rather than any traceable process. In contrast, multi-criteria analysis (MCA) uses explicit objectives and measurable criteria to assess the success of meeting objectives and has benefits over informal judgement decisions[10.12]:

- Both numerical methods and implementation are transparent and explicit
- Weightings may be applied, but are equally transparent and rational
- Measurements of performance can come from a variety of expert sources
- The objectives of the analysis can be easily changed to suit appropriateness
- An audit trail can be easily developed

The use of an MCA approach for the selection of materials is therefore preferable.

A variety of multi-criteria decision-making (MCDM) methods have been developed over the years, each with their own advantages and disadvantages. The methods include analytic hierarchy process (AHP), analytic network process (ANP), fuzzy analytic network process (FANP), criteria importance through inter-criteria correlation (CRITIC) and the technique for order preference by similarity to ideal solution (TOPSIS).

Multi-criteria decision-making processes require criteria weightings in order to calculate the preferred choice. Weighting is the numerical process that allows for the relative importance of a particular property or category to be shifted. The introduction of weighting may introduce judgement decisions about the relative importance, which is difficult to avoid unless objective weighting methods are used. There are multiple objective weighting methods, these are the mean weight method, entropy method, standard deviation method and the criteria importance through inter-criteria correlation. The mean weight method gives all criteria an equal weighting whereas the entropy method is based on the uncertainty of the criteria, with attributes with greater differences being allocated greater weights. The standard deviation approach is similar to that of the entropy method, as it allocates weights according to how large or small the variation is in a criterion. Whether subjective or objective, clear and consistent applications of weighting allow for auditing of the decision-maker choices. This should ultimately lead to a greater diversity and confidence in material selection.

10.6 QUESTIONS TO CONSIDER

To help improve confidence with new, innovative materials, asking questions about the material will help understand the material's behaviour. Any resulting property or performance can then be appropriately engineered. Some questions to ask of a new material are presented in Table 10.1.

Table 10.1: Questions to contemplate when considering new materials

Q1	What is its intended use and application?
Q2	What properties do I need to know?
Q3	How consistent are the properties?
Q4	What is its visual appearance?
Q5	Does the material have a preferred structural form?
Q6	What existing material is it like?
Q7	How is it made?
Q8	How could this material fail?
Q9	How friable is it?
Q10	Is any particular care necessary for storage/handling of the material in production or on-site?
Q11	What is the durability of the material in the temporary and permanent state?
Q12	Is it compatible or how does it behave with other proposed materials in the structure?
Q13	What is the impact of environmental conditions such as water and temperature?
Q14	What are its sustainable credentials; does it have an EPD?
Q15	Is there any rule-of-thumb design parameters to use for early feasibility design?
Q16	Are there any codes or standards to be used?
Q17	What is its comparable cost against other materials?
Q18	What approvals are required by the client/owner?
Q19	What are the programme implications of using this material?
Q20	How is it protected in the temporary situation on-site?
Q21	How should it be tested?
Q22	How many should be tested?

10.7 CONCLUSIONS

There is a huge variety of materials that are used within structural engineering and the selection of the material has impacts on analysis and design, construction, building operation and any conservation or end of life considerations.

Even with such a diverse canvas the industry is reliant on using existing materials. While understandable to ensure a high level of confidence in performance it has arguably negatively changed the perception of new materials. This is significant when the perception is that they don't conform or behave like existing materials. While information may be lacking, statistical methods can be used to quantify the relationship of performance. However, we have seen that many existing materials don't actually behave as we assumed, and embedded within design practice are assumptions and engineering approaches to deal with the challenges. For example, a material that shows no clear elastic and perfectly-plastic behaviour is akin to concrete, and those materials sensitive to moisture (or other conditions) are similar to timber which modifies the strength and stiffness. Ultimately the perception and confidence in a given material (from the designer as well as the contract specifier, the client, the financier and many others) will dictate its use; not necessarily the technical performance.

In Chapter 11 we will focus on physical testing as we consider how we might gather the data we need for new or uncodified materials.

REFERENCES

10.1 UK Green Building Council. *Embodied Ecological Impacts.* UKGBC, 2024. Available at: https://ukgbc.org/our-work/topics/embodied-ecological-impacts/ [Accessed: January 2025].

10.2 Expedition Engineering. *The Embodied Biodiversity Impacts of Construction Materials.* London: Expedition Engineering, 2024. Available at: https://expedition.uk.com/project/embodied-biodiversity-impacts-of-construction-materials/ [Accessed: January 2025].

10.3 Royal Institution of Chartered Surveyors. *Whole Life Carbon Assessment for the Built Environment (Second edition).* London: RICS, 2023.

10.4 Boote, S. and Lynes, A. 'Stone as a structural material. Part 1: mechanical properties'. *The structural engineer*, 98(3) March 2020, p6. Available at: https://doi.org/10.56330/VDXN2678 [Accessed: January 2025].
10.5 Obrien, E. J. *et al. Bridge Deck Analysis*. Boca Raton: CRC Press, Taylor and Francis Group, 2015.
10.6 Statistics Solutions. 'Understanding the Wilcoxon Signed Rank Test'. *Directory of Statistical Analyses*. Available at: https://www.statisticssolutions.com/free-resources/directory-of-statistical-analyses/how-to-conduct-the-wilcox-sign-test/#:~:text=The%20Wilcoxon%20signed%20rank%20test%20offers%20a%20robust%20alternative%20to,repeated%20measures%20or%20paired%20observations [Accessed: January 2025].
10.7 Walker, P. *et al. Rammed Earth: Design and Construction Guidelines*. Bracknell: IHS BRE Press, 2005.
10.8 Ministry of Housing, Communities and Local Government. *Publication of the Grenfell Tower Inquiry Phase 2 Report*. 2024. Available at: https://www.gov.uk/government/publications/publication-of-the-grenfell-tower-inqury-phase-2-report [Accessed: January 2025].
10.9 British Standards Institution. *BS EN 1995-1-1:2004+A2:2014: Eurocode 5: Design of Timber Structures - General. Common Rules and Rules for Buildings*. London: BSI, 2014.
10.10 The Institution of Structural Engineers. *How to calculate embodied carbon (Third edition)*. London: IStructE Ltd, 2025.
10.11 The Institution of Structural Engineers. *The regenerative structural engineer*. London: IStructE Ltd, 2024.
10.12 Dodgson, J. S. *et al. Multi-criteria Analysis: A Manual*. London: Department for Communities and Local Government, 2009. Available at: https://eprints.lse.ac.uk/12761/1/Multi-criteria_Analysis.pdf [Accessed: January 2025].

CHAPTER 11

Physical testing

James Norman and Scott Boote

11.1 INTRODUCTION

In Chapter 10 we looked in depth at classifying materials. While for some materials the properties are well-established, for others, as designers, there is more information required.

In this chapter we will therefore look at physical testing; from tests in the workplace to carrying out tests at a testing house. The chapter will look at a wide number of different considerations for testing and is built on the authors' knowledge of testing both in industry and academia and includes lessons learnt from experience.

It is important that testing should go hand-in-hand with analysis to confirm that parameters are acceptable and repeatable.

11.2 WHY PHYSICALLY TEST?

Physical testing can be a valuable mechanism at all stages of engineering design. When exploring initial design ideas, it can be useful to investigate different structural forms, or to demonstrate a complex behaviour to a less-experienced team member. During early-stage design we can use testing to confirm the behaviour of materials and existing structures, to understand better how they can be used efficiently. Testing may form the basis on which all design is completed (for example, for many proprietary products available on the market), and during the construction stage we can use testing to check our design requirements have been met 'in-the-field'. Even after a structure is complete, we can conduct further tests which allow lessons to be learnt for use in future designs.

> "*The value of a prototype is in the education it gives you, not in the code itself.*" Alan Cooper

Physical testing is both a hugely exciting and yet potentially fraught part of the process. Many engineers do not have much experience in this area, and yet it may also be the hands-on practical involvement which got them excited about engineering in the first place.

This chapter will start by considering the different facets of testing and builds upon some of the questions set out in Table 10.1. Why are we testing and what will be the outcomes? What do we already know? What type of test are we hoping to carry out? What is the scale of the test — are we looking at a single element or a whole building? Where do we intend to perform the test?

Before you proceed to read through the detail of this chapter, we would encourage all readers first to consult Annex C and D of *Eurocode — Basis of Structural Design*[11.1].

11.3 THE DIFFERENT FACETS OF TESTING

Physical testing is a large and complex field. Most projects will involve some form of physical testing, either directly or indirectly; from specifying cube tests for concrete to testing soil samples to calculate an allowable compressive strength. From using products that have been certified by testing, to using timber that has been graded through tests — a lot of testing occurs. However, as an engineer, to decide to test something outside the usual may be a big decision and one that requires a significant amount of thought. Before we explore how to commission testing and how to understand the results, it is first useful to understand the many facets of testing. These facets have all been labelled using different counting systems so that you can effectively class your own requirements.

Table 11.1 is a summary of all the different facets of the testing process. Over the course of this chapter we will explain each of these facets in more detail.

Table 11.1: Different facets of testing

Reason for testing*	What information is already available?	Type of test	What level of complexity?	Where can testing take place?
1. Idea exploration (TRL 1/2) 2. Proof of concept (TRL 3) 3. Product testing and grading (TRL 4/5) 4. Quality testing 5. Existing building 6. Academic study (TRL 1-5)	A. Full code B. None C. Academic studies/ papers D. Unpublished studies E. Existing buildings F. Failure case studies	a. Strength/ stress/strain b. Fatigue/ cyclic tests c. Dimension/ density variation d. Fire e. Other material properties f. Existing buildings g. Buildability h. Long-term load effects j. Other types	i. Material ii. Bond/ connection between materials iii. System iv. Whole building	v. In your garage/ shed/office/ workshop/ at your desk w. On-site/in the factory x. At a university y. At a testing house z. Existing buildings

* TRL = Technology readiness level

11.4 PURPOSE OF TESTING

There are many reasons for testing. Some of the most typical are discussed in Sections 1–6. If you are considering testing, there is normally a good reason for this. Knowing what you are hoping to find out is important. For most practising engineers you will either be doing an initial study, a basic study, a test of an existing structure or tests for quality control purposes. However, we have included product testing and academic studies as well, so you are aware of the breadth and purpose of the different types of studies. The purpose of testing could also be benchmarked against 'technology readiness levels' (TRLs)[11.2]. While there is not a straight one-to-one correlation, by exploring the concept of TRLs it may also help engineers ask questions they wouldn't normally think to ask. The levels have been shown in brackets in Table 11.1 as a guide, where appropriate.

1. Idea exploration

Often on a project it can be useful to have a grasp of the way a material behaves or how a system functions. To understand this, we may want to do an initial study. The purpose of this study is not to have an exact understanding of a material property, we are not looking for a mean and standard deviation, but instead we may be looking to establish more basic answers on the properties. Will this be a brittle or ductile failure? Does this material behave in a similar way to another material? Does the structure have issues with instability? Is the material isotropic or anisotropic? The work done at this stage is often useful because:

- It may enable us to rule out a solution — this may save time and money on further testing
- It may enable us to understand better what questions we want to ask at later stages of the design process, or when doing more precise testing. For example, if a material is brittle at failure we will need to make sure we apply load at a suitable rate so we can observe the

failure and have a rough idea of when this might occur, so we instrument appropriately

- It may provide insight into how a system works. On the Tate Modern facade, long before any instrumented testing was carried out, a stack of timber blocks was set up with varying staggers (Figure 11.1). This showed that above a certain angle the blocks require a top plate to thrust against to enable arching action. Although this could have been predicted with a mathematical model, knowing what the solution looks like helps in defining the analytical model, hence simplifying the process
- These tests can contribute to the engineer's level of understanding at an intuitive level. Visualising the behaviour or failure mechanism of a system can inform the next stages of design or testing, reduce unconscious bias or confirm/disprove preliminary assumptions
- Tests conducted at this stage can be fast and inexpensive allowing design teams to identify the best options

Figure 11.1: Initial study for the Tate Modern facade

While some form of measurement is useful, especially if you are carrying out an initial study which will inform a more detailed study later, it is not always necessary. The stack of timber bricks provided useful insight without needing to measure anything or add instrumentation. However, quantifying outputs is often much easier than people imagine and can be very powerful.

2. Proof of concept

This is the type of study practising engineers are most frequently likely to consider. It will involve looking at a small subset of characteristics, behaviours or performance. The aim will be to collect enough data to complete the design with confidence, but not so much as to produce an academic paper, information for a code of practice or for more generic product testing.

The types of tests are covered in Section 11.6. It is likely that you will not be considering all of these but instead focusing on some specific information that you require for the design, such as the stress-strain behaviour or the behaviour of a specific, pre-selected, component (Figure 11.2).

Figure 11.2: Component testing for a bespoke project

3. Product testing and grading (BRE, BBA, CE marking, industry specific, etc.)
We are not suggesting that you use this book for large-scale product testing, but it is good to know that this occurs, that there are ISOs/BS ENs for this and that there are quality requirements. There is a legal requirement that all products manufactured to a code of practice are tested and have CE marking. Currently in the UK, all construction products must comply with the construction products regulations. Under the Building Safety Act, new product regulations will also impose a general safety requirement on all construction products. When product testing occurs, there will be a more systemic approach to testing, with multiple repetitions to ensure a higher level of certainty around the outcome. While much of the data for large-scale testing is often unpublished for commercial reasons, some helpful information may be available from product specifications and testing certificates. For example, if you want to do your own testing of a similar product you may be able to use the same testing standards and even the published design data, if you are applying a Bayesian approach (Chapter 15).

A subset of this approach is the non-destructive grading of products, such as timber. Grading has specific codes which establish not just the type of test, but also the number.

Australian and New Zealand codes on grading timber are particularly helpful if you can access them, as they not only provide clear guidance on how to grade and the number of tests necessary, they also include worked examples that can be followed[11.3,11.4].

4. Quality testing
Most of us will engage in quality testing in our careers. This could take the form of a pile load test or concrete cube test, for example. We have a design and an expected outcome, and we are now checking the outcome. A small number of

samples is sufficient as we already have a parent data set (for example, for a concrete mix we know that for x amount of cement, y amount of aggregate and z amount of water we will get a set strength with a standard deviation) so all we are doing is adding to the data set. We don't need a mean and standard deviation for our test data — which is why we can often simply take a single reading at seven days and another at 28 days with one cube spare.

However, we could also reimagine quality testing. What if we designed a building for a set loading, with no future flexibility planned, based on the current statistical analysis? If we then carried out a more detailed set of quality tests, we could increase the design load (assuming the results were favourable) as long as we know the governing failure and load combination, rather than carry out any strengthening work. We could do it at the point of construction, rather than 30 years later. This is normally included in the construction phase plan as an inspections and test plan (ITP) and the results would be included in the health and safety file that is passed over to the client/ building owner upon completion of construction.

5. Existing building

If a building has already been built, then we may also want to do some testing. This may be to understand what was built if we don't have drawings (or we are not confident that what was drawn reflects final construction), to look for signs of degradation or to increase capacity by finding out a more representative value for the material property. Testing can be small-scale, for example crushing cored out cylinders, or large-scale, in testing whole floors under load (Figure 11.3). There are some specialist companies that provide testing of existing buildings. More information on approaching investigating existing buildings is discussed in Part 4.

Figure 11.3: Load testing an existing building

6. Academic study

Finally, academic institutions may do testing to understand something better. This is often more thorough than proof of concept testing, but not as rigorous as product testing. The focus may be on not just getting a single value, but understanding how something changes when exploring different variables. For example in connections in bamboo we may vary dowel diameter, material thickness, fixing spacing, fixing edge distance, etc. There will often be an analytical or numerical (e.g. FEA) model that the experiments are being compared to, which will predict how these variables change the response, and the experiments will be designed to verify these. As we will see later, these studies can be very helpful in understanding the problem. The tests themselves will also, if possible, be carried out to codes of practice when available, to ensure repeatability.

For example, for testing the performance of dowels in bamboo, while no codes are directly applicable it would seem appropriate to apply the timber codes, BS EN 12512:2001[11.5] and ASTM D 5764-97a[11.6], as both materials are natural, fibrous and anisotropic (Figure 11.4)[11.7].

When looking at academic studies, as well as considering data it is also useful to refer to the methods and testing standards applied, as these may be helpful in creating repeatable tests for our own specimen.

Figure 11.4: Testing a dowel connection in bamboo

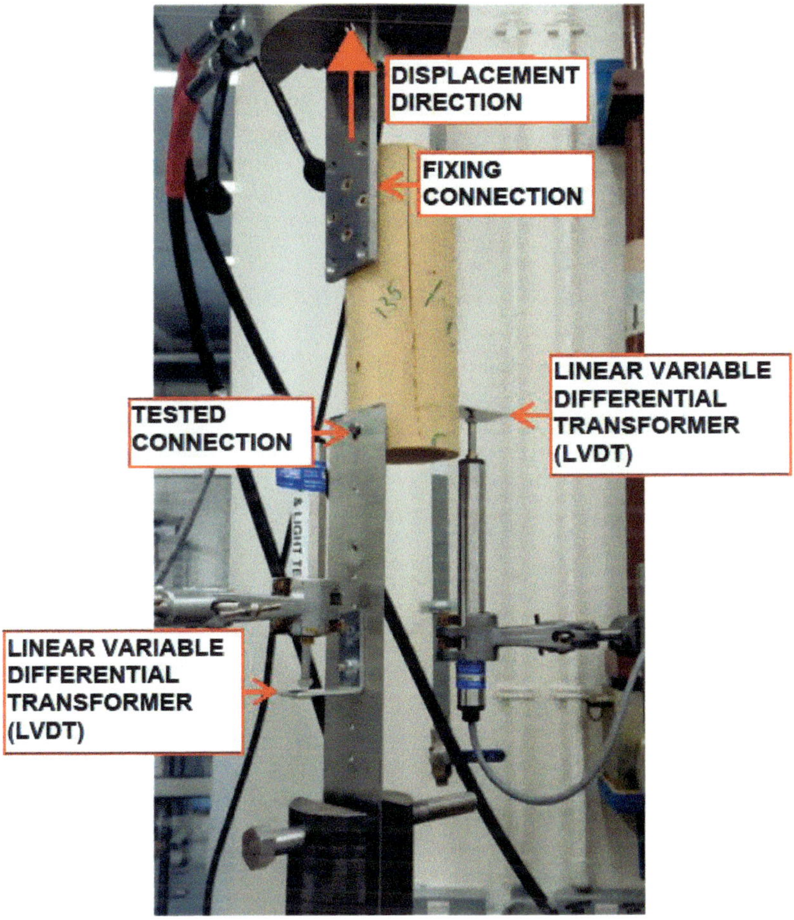

11.5 WHAT INFORMATION IS ALREADY AVAILABLE?

Before we start to plan our testing in detail it is important to understand what information is already available. The more you know, the better the test and smaller the uncertainty when we come to the design stage. It may even be that we don't need to carry out tests that we first imagined, as someone else has already completed them.

A. Full code

As engineers we love codes — they make us feel safe and with good reason. The information contained in them is based on both a variety of tests and significant experience. Codes also include long reference lists.

If a code is available, perhaps we don't need to do any testing at all. However, there can be good cause for testing even if we have information at a codified level, for example, for the purposes of quality assurance, or to check an existing structure. We may consider testing to enhance what we already know from the code. For example, by testing the statistical distribution of dimensional variance in a prefabricated element we may be able to reduce the partial safety factor on the self-weight.

Finally, codes are produced across the world and contain valuable information. Casting the net more widely increases the chances of finding a code which covers the material in which we are interested.

B. None

I have included this category as a provocation. Often engineers believe that there is no information, but I can't think of a single case where we know nothing about a material, structural form or loading. What we know may be very broad (I know my material is weaker than concrete, but stronger than custard) but this is knowing something.

We also know that the strength is positive (a negative strength makes no rational sense for a material). So, we may not know much, we may not have a number for the strength, but we might have a wide range. This information, while not as useful as more precise data, may still be helpful, even if it is just to aid in the design of a test, or to check the outcome makes sense. Importantly, we should know what key questions to ask, as outlined in Section 10.6.

Some information can come from having an intuitive understanding of the world. When considering the design of a structure in an unfamiliar material we can make assumptions based on similar materials. For example, for rammed earth based on very limited experience we can assume that the material will behave better in compression than in tension, have low (relative) compressive strength but high (relative) stiffness. These assumptions can be used as a basis on which to direct further investigation and research.

C. Academic studies/papers

Academic studies can be rich in data. They may not answer the specific question we are considering, but they will provide some data. This may be aggregated data (plots and tables) hopefully with a standard deviation. If the study was recent the full data set may also be available. The study may provide other valuable information — which codes were used for testing, which materials this new material behaves like, what failure modes are most likely?

Also, behind every academic paper is someone who carried out the work. If you can contact them (and bearing in mind the study may have been carried out some years before) they will be a mine of information. They may have information on buildability, challenges and constraints that often don't make it into the paper. Often studies are limited by time and resource and finding out why decisions were made about the scope of the study may also provide information on the study itself.

Academics are looking to publish new information. If a study looks like it may not yield anything new, the direction of study might change, and the focus move to another feature. This lack of 'new' in and of itself may be very useful — for example, it may lead to another paper, another academic who did more work in the area in which you are interested.

Also try to locate PhD theses, MSc theses or other reports. These will be harder to access and you need to be mindful of the level of academic review (for example, a PhD goes through the *viva voce* process and is resubmitted until the assessor is satisfied, but an MSc is marked but never corrected). Some papers will have come from a much larger study and more data and more information on how the study was carried out may be available there. There may be a more thorough literature review and analysis of different testing methods. You may also find a much more comprehensive explanation of the limits of the study and what the next steps should be (but may have never been carried out). These should be available from university libraries and should now also be available electronically. Reports summarising grant outcomes may also hold more information.

D. Unpublished studies

Not all research is published. There are a large variety of reasons for this including:

- Time — publishing a journal paper takes time and often people run out of that. We know this from personal experience. Having left academia for industry, much of my PhD thesis never made it into journals. Even when some findings are published there may not be time to share everything discovered

- Level — all universities will carry out undergraduate research projects and many will also do postgraduate projects. These will contain data. Much of it will be less reliable than other testing data, but more reliable than no data at all. Some universities will keep a store of projects — but finding them may prove challenging
- Intellectual property — some universities and some companies will, for reasons of intellectual property or security, choose not to share data. If you pay a university to carry out a test, you need to consider if you want that information shared more widely. On the other hand, when planning to carry out a test, information may be available from other companies that have carried out tests in the past. It may be possible to involve them in the project, either by agreeing to share data/results of any testing you do in exchange for data they have, or by appointing them to the project, possibly as a reviewer

There is a lot of information out there. At some point you will need to decide if the effort required to find it is worth the investment, or if it would be more efficient to perform the tests yourself. However, don't assume that, just because the data hasn't been published, it doesn't exist. Talking to peers and colleagues is a great way of understanding what has been done before, and where gaps in knowledge exist. Having a network of sources across the industry can also prove beneficial. With networking sites like *LinkedIn* it is possible to reach out to a large number of practitioners from across the world.

Figure 11.5 shows an undergraduate test on a rammed earth wall under seismic condition, the data from which was never published as an academic study and Figure 11.6 shows testing of pressed bricks, using traditional methods and is an example of on-site testing.

Figure 11.5: Undergraduates testing a rammed earth wall under seismic conditions

Figure 11.6: Site-testing pressed bricks using traditional methods

E. Existing buildings

If a material has been used in construction, we already know something about it. We know it can be used to build with. We can start to estimate the strength (upper and lower limit), and we can start to observe long-term performance. We will obviously know a lot less than we know from carrying out detailed testing, but we still know something. This can be true for any type of construction. In the UK if a building has been built, someone somewhere must know something about it! Building control may have been involved, designs will have occurred, even if these are based on the crudest of tests. If we are thinking about a traditional material that has not been 'designed' or vernacular construction, then the test is in the existence of the building itself. This can still teach us a lot about the way the material works. Are there signs of long-term movement? Do the elements exhibit tension cracking? Are there signs of crushing at points of high-bearing stress?

Often, engineers rely on the unwritten '100-year rule'. If a structure is older than 100 years and does not show signs of significant distress, it can often be deemed to have undergone a full load test and be 'fit for purpose'. Significant caution needs to be exercised if this approach is to be adopted. It is often the case that an engineer's involvement will only occur if changes are to be made to the building. These changes may render the previous 100 years irrelevant. It is important to understand what aspects of the structure or its use are to be changed, and how this may affect the remaining elements. For example, if the changes affect only the floor plates and the stability system remains unaltered, it may be safe to assume the building can withstand any wind loads it will be subjected to, based on historic performance. Conversely, if the purpose of a building is to be changed from residential to retail, for example, the structure may not be capable of resisting the additional design floor loads. In this case further analysis and testing would be required.

F. Failure case studies

Sometimes things go wrong and these incidents can teach us a lot. Some failures redefine codes, others are less dramatic but are still concerning and are captured in CROSS (Collaborative Reporting for Safer Structures) reports[11.8]. A key lesson from these incidents, is that failure often comes from an unexpected source. Generally, the root cause is not inadequate element capacity or inadequate analysis (although both of these can occur) but another mechanism entirely.

From poor cleaning of bond surfaces when using adhesives, to assuming a pin when one doesn't occur, failures can reveal all sorts of information about what, where, how and when we should be testing. Figure 11.7 shows a failed resin anchor that occurred due to lack of hole preparation.

Figure 11.7: Failed resin anchor

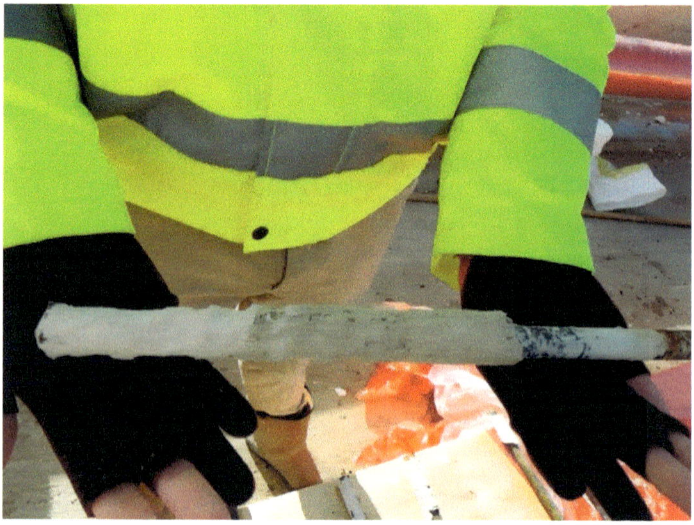

Failures can also suggest different conditions in which we should be testing. What were the environmental conditions during installation, operation and at the point of failure? There are lessons to be learnt when things go wrong, even if we have data to suggest it should work, just as when structures are evidently working, but we don't have data to prove how.

All this information will be important both in the design of a test (e.g., are we replicating lab conditions or site conditions when checking bond strength?) and in working out which tests to do (what is the key design data we need?). A search for previous failures in addition to existing data will often yield extremely useful results.

11.6 TYPE OF TEST

There is a wide variety of different types of tests. Even if we are trying to test something as simple as strength, there are many ways to do this. It is not possible to list all of them, but by referring to some of the most common we aim to highlight the range of tests available, and what you might hope to discover with each.

a. Strength/stress/strain

Most engineers, when they consider testing, think of crushing cubes and failing samples in tension. Strength tests can vary significantly in scale, from pulling material coupons in tension to establish their stress-strain behaviour, to load-testing piles. At the heart of all these tests are strength (or stress) and displacement (strain).

Tensile strength tests

For testing the tensile strength of most materials, a coupon is typically used. A coupon is a test specimen of known dimensions, like length and cross-sectional area which is 'gripped' by a machine at both ends. We then apply a tensile force/strain to the specimen. Coupons are designed so that the area at the two ends is larger than in the middle. This way we can ensure that the failure is at a preferred known location, and that this location can be observed and/or instrumented. For some materials a coupon won't be practical, such as stone. In this case a four-point bending test can be used to create constant flexural tension along a predetermined length of the specimen (Figure 11.8).

Figure 11.8: Three- (left) and four- (right) point bending

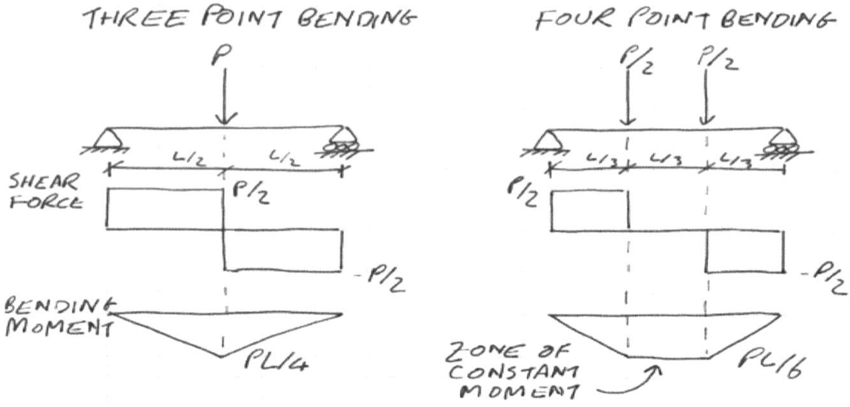

Compressive strength tests

For compression we can't use coupons as they may buckle, so instead cylinders or cubes are preferred. However, there is also a problem, as for both cylinders and cubes the mechanism and load at failure is affected by the constraint provided by the two loading faces. In concrete design, we then apply a factor to get from the confined stress to the unconfined stress because of this behaviour. This highlights how the data obtained from a test must be carefully processed and interpreted.

Other types of strength test

Testing machines/mechanisms come in a variety of forms, from applying static load in the form of kentledge or masses, to using jacks to apply load, either using a hand pump or powered hydraulic pump (Figure 11.9).

Figure 11.9: A timber flitch beam loaded with masses (top) and a rammed earth wall loaded horizontally with a hand-pumped jack (bottom)

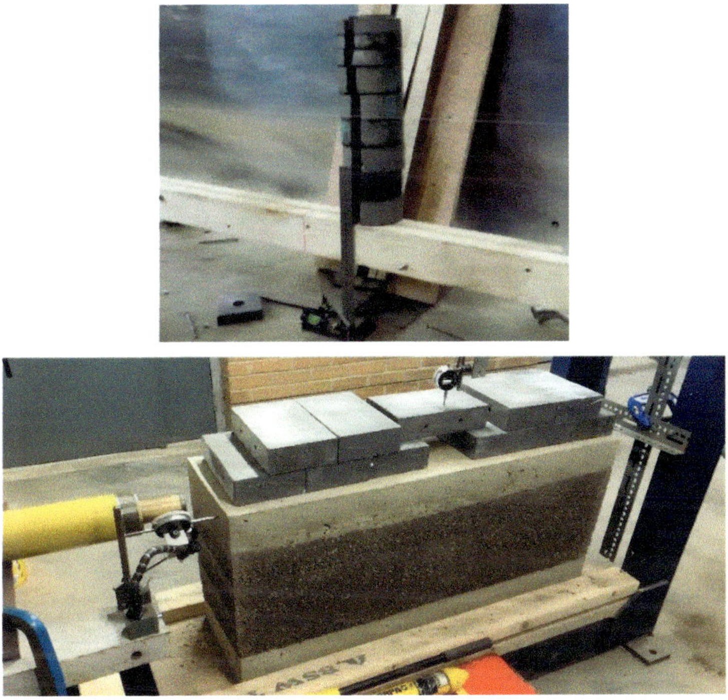

With some of these loading methods we have a choice as to whether to control the applied load or displacement. Controlling load provides us with a high level of certainty that we apply the load in the desired timeframe, but also means that we get little to no post-failure information. Once the peak has been reached the loading continues to increase and the sample will fail. This is the case if you are applying load as weights, for example.

Understanding the post-failure behaviour can be very beneficial however, and therefore, if possible, using displacement to control the test can be useful — we can apply a known displacement every second and then, as the specimen goes past its ultimate load, the testing machine will start reducing the load to achieve constant displacement until the specimen has failed. This requires advanced control, and sometimes the loss of strength is so rapid it may not be possible to gather this data, even if we want to (Figure 11.10).

Figure 11.10: Load controlled (left) and deflection controlled (right) experiment with the same specimen

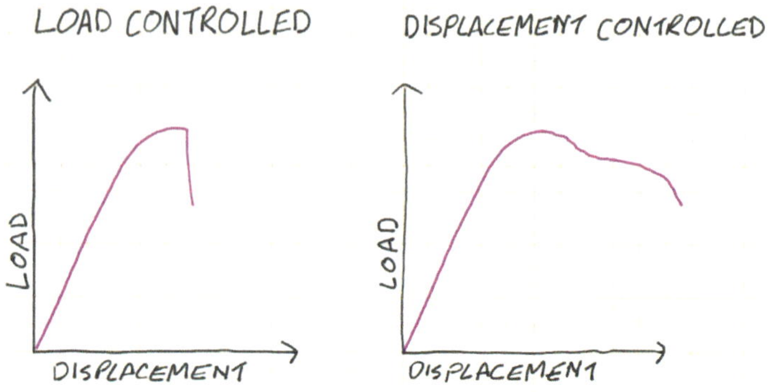

b. Fatigue/cyclic tests

Further to understanding the strength (stress-strain) behaviour of our material, we may also want to understand its fatigue or cyclic behaviour. To determine whether we need this information we may want to consider what the stress variation is likely to be. Causes of significant changes in stress include, but are not limited to:

- Earthquakes
- Materials/connections subjected to horizontal (wind) loads which can occur in both directions

- If the self-weight/permanent action is small and the variable action is, as the name suggests, very variable (for example, bridges) a lightweight roof subjected to load reversal under wind load, or a lightweight floor subjected to variable crowd loading
- Structures that are exposed to large temperature variation or other environmental effects

Cyclic tests look at the behaviour as you apply cycles of displacement with predetermined increases in magnitude, so for example in BS EN 12512:2001[11.5] the test method specifies three repetitions of each displacement cycle increasing at 20% increments up to (and beyond) 100% of the failure displacement. In fatigue testing the number of load cycles will generally be in the 1,000s or 10,000s at lower load than the failure. These tests can take months and are therefore generally carried out as research.

c. Dimension/density variation

One of the unknowns in design is variability of the structural element. This impacts on both the resistance of the element (a column cast smaller than intended will have a higher stress) and also on the load. If a floor slab is 10% thicker than intended the self-weight will be 10% greater. The geometrical variability of individual elements is generally not considered in the codes (although the masonry code — BS EN 1996[11.9] — does allow for it in the application of the partial safety factor where quality assurance has or hasn't been applied) and is part of the material uncertainty. The variability in load is considered, and accounted for by the partial safety factor, but this is generally consistent across all construction types.

Recent developments in technology and the move towards prefabrication allow greater levels of certainty in this area. The use of prefabricated elements, particularly where using some means of digital automation (CAD/CAM/CNC) results in a much higher level of control. These methods make the mean and standard deviation of the product's dimensions and density much more readily available and can be used to reduce design loads. Likewise, point-cloud surveys provide very accurate dimensional data. This information can be incorporated into load calculations and analysis of existing structures.

If these modern methods are not available, there may be other methods that can be deployed. If you are working with a precast manufacturer, they may be able to supply dimensional and density data. Or if you are assessing an existing slab, the cores that you take can be used to provide not only strength data, but also the slab thickness and material density.

We can also see variation in materials both depending on the axis the load is applied in (if anisotropic) as well as within the material itself. For example, timber strength changes significantly when a knot is present. Stone is often made up of layers and failure occurs at the location of flaws[11.10].

Finally, we should consider spatial variation at a larger scale. How variable are rib spacings? How much variation is there in formwork, or variation of void formers? All of these can affect both the overall self-weight and also the stresses exhibited in an element.

d. Fire performance

When we are not providing fire protection to our structure we should be carrying out fire testing (Figure 11.11). This should consider both the combustibility and performance of the material at elevated temperatures (Section 10.4.4). It is worth noting that there can be a significant difference between test data and performance *in situ*. If fire is a concern the advice of a fire engineer should be sought before proceeding.

Figure 11.11: Fire test of a Modcell panel

e. Other material properties

As highlighted in Section 10.4, beyond the stress-strain behaviour of a material there are many other types of information that we might want to gather. These include:

- Isotropic or anisotropic
- Coefficient of thermal expansion — this is particularly important for composites where variation in thermal expansion will lead to additional stresses and possible distortion
- Shrinkage — both initial drying shrinkage and changes in dimension due to changes in moisture content. Again, this is important for composites and also when considering location and size of movement joints
- Performance at different temperatures — especially considering performance in fire
- Impact toughness — this is typically calculated using a Charpy impact test
- Hardness — some materials (such as unfired earth blocks) are not particularly resistant to surface wearing.

Understanding the performance, especially where surfaces are exposed, should be considered

- Chemical composition and whether the material is likely to react with another material (or act as a catalyst) or with the air. Examples include rusting, changes in pH which lead to corrosion in reinforced concrete and bimetallic (galvanic) corrosion
- Following on from this, is the material organic and can it rot?
- Are there any other forms of long-term degradation? For example, bamboo is susceptible to being eaten by beetles and various types of mite
- Thermal conductivity — useful in the design of facades when calculating u-values and for thermal mass
- Antiseptic — especially if a material is to be exposed. How do pathogens behave on the surface?
- Embodied carbon — as the aim of using new materials is often to reduce carbon it is important to understand embodied carbon, and consider how it might reduce as production is scaled up
- Sensory concerns, including smell, visual appearance and tactile performance

This list is not exhaustive and for any material it is important to try to understand all the different facets, work out what is known, and also work out how important it is to know. The aim is to reduce the unknown unknowns to an acceptable level, while accepting we will never know everything about a material.

f. Existing buildings

Existing buildings require a whole host of other test types. From loading a floor and measuring the displacement or strain (we don't want to test it to failure — Figure 11.3) to using a reaction mass shaker or an eccentric mass exciter to excite the building and check the vibration response (Figure 11.12).

Figure 11.12: Testing the floor of a building with a vibration exciter

We can also instrument existing buildings without running any 'experiments' to assess them. From measuring the movement of a building (for example, where we see cracks) to adding an accelerometer and measuring ambient vibrations. This is a growing area that could be important to give us a much better understanding of buildings and structures in the future. For example, we could use this data to manage building performance, from controlling load to ensure we remain within serviceability requirements to limiting movement to control vibrations during an experiment. If we collect data from a number of buildings with instrumented floors, we could provide a load history demonstrating a more evidenced-based loading for that loading type. With the advent of the internet of things (IOT) we are currently looking at large-scale instrumentation of existing infrastructure at very modest costs and current satellite data transfer is available at around £30 per annum for remote locations. All of these approaches could be powerful in reducing embodied carbon when assessing existing buildings.

g. Buildability

One of the most important tests that you carry out will probably happen without you realising it. Understanding how a material behaves as you make it, how workable it is, whether it has any quirks during construction, is really important. For example, one might assume lime concrete behaves just like concrete, except it doesn't. It is much more 'sticky', it is less workable, is harder to place and harder to compact. By making specimens, even without testing, we discover this. Furthermore, this is the sort of information that is often overlooked in academic papers, but may be mentioned in a thesis, and should be raised if you speak to someone with prior experience of the material.

Buildability also goes beyond individual materials. The appearance of exposed structure may also be of interest, understanding how a system is put together and how it will look (as well as how it will behave) may be a consideration. A typical example is in exposed concrete where, as a minimum, you may want to limit surface pore sizes and avoid honeycombing. You may also consider the surface appearance (by texturing formwork) and how joints appear. Much of this will require samples to be constructed.

Beyond buildability it is also important to consider reusability; can the materials be reused without losing its materiality? For example, bricks in pre-WWII houses can be cleaned of lime mortar and reused. Those in post-war houses built of cement mortar cannot be easily cleaned and reused as bricks. In making and then dismantling our specimens we will quickly learn how easy it is to reuse.

h. Long-term load effects

Long-term load effects are very difficult to measure. They require tests that last for long periods of time which are both expensive and use up space. As a result, gathering this data can be hard but it can also be important.

There are two different dimensions to consider. Firstly, does the structure creep with time? This will affect displacement. Secondly, is the strength affected by the load duration? The most obvious example of this is timber, where we account for the load duration in the design.

While formal tests may be challenging there are some things you can do. Firstly, look for examples of buildings made from the same materials. Do they have signs of long-term movement? A basic understanding of long-term movement in timber can be gained simply by looking at the ridge line of historic timber roof structures (Figure 11.13).

Figure 11.13: An existing timber frame building — the bow of the roof suggests that the timber has experienced creep over time

Secondly, can you construct and monitor a sample, at least for the duration of the project? Could you create a 'bench' out of the material and monitor displacement or strain over time (Figure 11.14)? While a year's data is not as useful as 50 years it is a lot more useful than no data at all and may help predict the long-term performance of the material.

Figure 11.14: Lime concrete bench

Existing buildings can often be considered as a long-term load test. For example, conducting a thorough visual condition survey of an existing building and looking for evidence of long-term movement could allow an approximation of the long-term behaviour of the underlying ground. An assessment of the long-term loading can be made, along with investigations to determine the foundation type, geometry, etc. Using this data, an approximate allowable bearing pressure can be estimated and used in the design of new foundations.

j. Other types

The list of types of tests is clearly very long. Beyond all the tests detailed previously, other tests include:

- Wind tunnel testing — to understand the wind load on specific cladding panels (Figure 11.15), structures (Figure 11.16) and the external environment for pedestrians
- Facade testing — to certify that the facade remains watertight, airtight and to check thermal performance

Figure 11.15: Cladding panel wind tunnel test

Figure 11.16: Dubai EXPO sunshade wind tunnel model

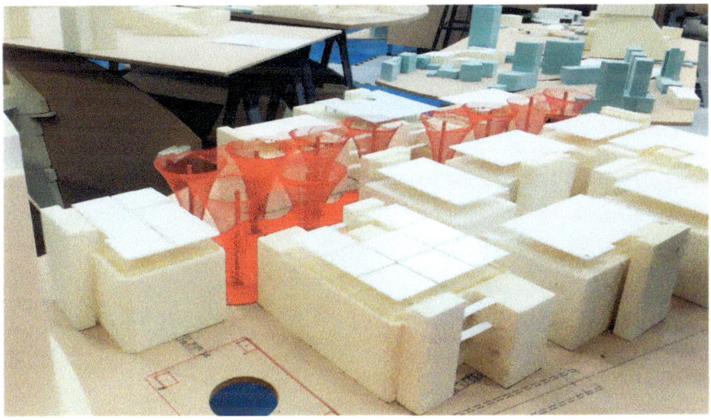

- Passivhaus testing — to check the degree of airtightness of the building[11.11]
- Acoustic performance
- Energy performance

11.7 WHAT LEVEL OF COMPLEXITY?

i. Material

When most people think of testing, they often think of learning about a new material or better understanding a known one. For engineers this is often assumed to mean understanding the stress-strain relationship, although, as outlined previously, there is a wide variety of different properties that we may need to understand. Material tests are familiar in the form of cube test and testing of site soils. Generally, they are small so we can fairly cheaply carry out a few to verify design parameters and improve the degree of certainty (Figure 11.17).

Figure 11.17: Testing concrete cube (left) and rammed earth cylinder (right)

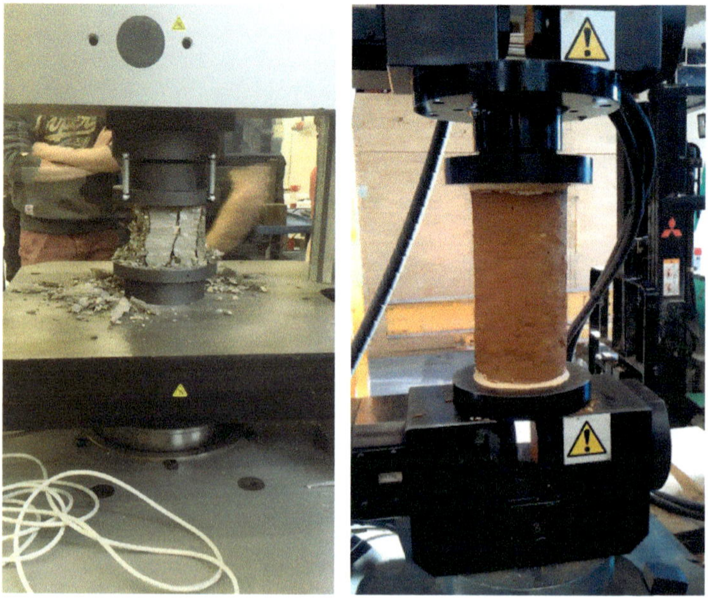

ii. Bond/connection between materials

Once we understand how an individual material works, we need to consider how it interacts with other materials. This includes, but is not limited to:

- Understanding the strength of gluing/bonding components — for example, finger gluing sawn timber elements
- Understanding the strength and stiffness of connections with fasteners in the form of screws, nails, bolts and dowels — for example, screw fasteners in bamboo (Figure 11.4)

Combining materials often increases the level of testing complexity. We need to be clear what it is we are testing and why. We also need to be very careful extrapolating data, as a failure mechanism that we might see in one sample may change if we alter the parameters of the sample. For example, in timber/dowel connections the failure type changes as a function of the material stiffness and dimensions. This is modelled with the Johansson equations[11.12] which approximate (but don't fully capture) the behaviour of timber dowel interaction. If the thickness increases, this can change the failure type and hence load carrying capacity (Figure 11.18)[11.13].

Figure 11.18: Different failure types in a timber connection with a dowel fixing

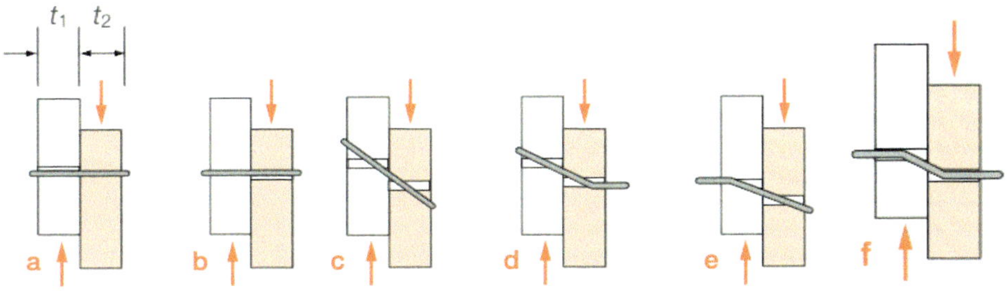

iii. System

Once we understand materials and the way these materials interact with fixings or how bonds occur, we may still want to understand how a full system works. This may involve testing the entire system or just a section of it. At this stage the aim is not to try to understand the way individual components behave, we have hopefully established this through earlier testing and analytical modelling, instead it is to confirm the validity of the assumptions made in devising these models.

Examples (Figure 11.19) include:

- Testing a timber concrete composite (TCC) floor
- Testing a brick and stainless steel perforated facade for the Tate Modern
- Testing the post failure integrity of a glass floorplate for a glass footbridge at Oxford Brookes University

System testing can also occur when we are assessing existing buildings. These are far more complex, but if we are load-testing or vibration-testing an existing floor we are essentially testing a whole system.

Figure 11.19: (Left) Full-scale testing of a timber concrete composite floor under a point load. (Middle) Load-testing the Tate Modern brick facade for horizontal actions. (Right) Glass bridges which required post-failure integrity testing before being constructed

iv. Whole building

Unlike other engineering disciplines structural engineers are typically unable to build whole buildings at full scale and test them. That is why every building is a prototype, and is why we spend so much time creating numerical models to try and estimate how the building will behave. But is this true?

The reason we can't build whole building models is because of the incredible expense. But what if we built smaller versions? Well, we need to be very careful with scaling. In simple terms axial stresses scale linearly, so if we made a half-size model with the same material density and properties, the stresses would remain the same. Bending stresses scale quadratically. For this reason, the likes of Frei Otto and Antoni Gaudí were able to make hanging chain models and tensegrity models of tension and compression structures long before the use of FEA, and could scale up the results and expect the same outcome (Figure 11.20). If you tried to do the same thing with bending structures (which almost all buildings are) the results would be catastrophic.

Figure 11.20 Measurement models were used by Frei Otto to develop his designs

In simple terms, scale models which emulate bending structures can only really be used to check/verify behaviour. Is the deflected shape as you'd expect? Do the initial signs of failure (cracking, spalling, etc) occur in the locations you had predicted? This type of model can be extremely useful but must be used with a large degree of caution. If the load path is complex, and is affected by the relative stiffness of different elements, you may observe significant differences in behaviour between small-scale models, and full-scale structures.

That being said, it is possible to make scale models, but we need to think carefully about the scaling. For example, experimental earthquake models are scaled to enable them to fit on a shake table. This requires careful consideration and a high level of expertise. The outputs are generally only used for research and are benchmarked against other models, both physical and numerical. The challenges around scale models are not limited to modelling them, but also creating the environmental and operational conditions which may affect how they perform (Figure 11.21).

Figure 11.21: Scale model of viaduct bridge on a series of shaking tables to model the effects of spatial variation of ground movements

In addition, once a building is built it can then be tested. This already happens in the form of post-occupancy evaluation and behaviour monitoring of existing buildings. As it becomes easier and more cost effective to instrument buildings and analyse the data, the idea of monitoring buildings is becoming more prevalent. We may in fact see a time when a building is instrumented and this data is used to alert the user to issues such as overloading as a way of controlling deflection. Outside of buildings, structural health monitoring has also been used to monitor the behaviour of bridges and other critical infrastructure.

11.8 WHERE CAN TESTING TAKE PLACE?

Testing can occur in a large number of different places. Determining where is best to carry out your test will depend on what you are trying to discover, how accurate you need to be, what level of expertise you require and what equipment you are able to access.

v. In your garage/shed/office/workshop/at your desk

Some of the world's most important discoveries have been made in a makeshift workshop or shed. Famously, Sir James Dyson built 5,127 prototypes in the creation of the world's first bagless vacuum cleaner[11.14]. Many of these were made from plastic and cardboard and were intended to test concepts. This approach can be adopted, and is often very informative, when trying to understand structural behaviour.

You will, almost certainly, want to do some testing in your own informal context. This may include in your office, workshop, garage, shed or even at your desk! This type of testing typically aligns with carrying out an initial study. From holding a piece of the material and asking if it is stronger than custard, or weaker than concrete, to doing some initial load tests, it is amazing how much you can discover with a few simple tools, a risk assessment (which is essential in all environments but even more so in this one), and a computer for data logging (micro computers such as an arduino/ raspberry pi/pico pi[11.15] are a cost-effective way of doing this).

By carrying out this initial testing you can develop a simple understanding of how a material behaves. While data is helpful, don't underestimate how useful the observations you might make can also be, from types of failure to simple features of structural behaviour. Is it stiff or flexible, brittle or malleable? Does it return to its original shape or stay in its new shape when you deform it? All of this will help you understand how to model it going forward and much of this can be observed as well as measured.

Even if you plan to go on to do basic testing, having some prior understanding will really help.

In addition, simple models can help us understand complex structures. We may have an FEA analysis and want to know — does this really deflect like that? Commercially available products such as Mola Structural model kits[11.16] could assist here, or a simple paper or cardboard model may be enough to persuade or dissuade us that the computer model is performing as we might expect (Figure 11.22).

Figure 11.22: Helical stair cardboard model

w. On-site/in the factory

In the factory a lot of data is already gathered — it may be helpful to gather as much as possible. In addition, on-site, there are already many tests that occur, such as pile tests, slump tests, taking cubes. In addition to this there might be buildability tests, and you may be able to use a small portion of the site to carry out some longer term tests. The tools you use will be similar to those mentioned previously but you are typically now outside and on a live building site. This brings some advantages — people will be less worried about the mess, but also some other challenges including protecting samples if they are required for a long duration and the additional health and safety requirements of a working site.

Examples include, but are not limited to:

- Measuring durability (and the resistance to erosion and weathering)
- Monitoring long-term movement and creep and understanding the buildability
- Speed of construction
- Necessary equipment
- Training
- Consistency and stickiness (if the material is easy to use/place)
- What specialist skills may be necessary for its construction

Even for traditional materials such as concrete, samples may be constructed to check colour, finish, formwork and detailing where the concrete is to be exposed.

x. At a university

Universities are well set up to carry out experiments and will have access to testing machines and measuring tools which are outside most practices' budgets. They also have expertise in testing and may also know more about the material you are interested in.

There are two ways to approach a university. One way is to ask them to do testing on a commercial basis. This will involve a specification for the works, a clear agreement on who is doing what and who is supplying what (they may expect you to make samples, for example). They should be able to fit in the testing in a reasonable timeframe, although many universities will have busy periods where they prioritise student projects, making testing in these times harder. Depending on what you are testing you may also want to commission some initial data analysis, so you don't just get a csv file with millions of data points and no explanation. You will also need to agree intellectual property. Are the results just yours or can the university also share them? This will partly depend on the novelty of the project; the more novel, the more likely an academic will want to publish them. This will also impact how keen a university is to do the work. While all testing is possible, it is more appealing for an academic (and their researchers) to work on something novel that can lead to publications and new learning. This will increase their appetite to work on the project.

The second way to approach a university is to partner with them. This could be in the form of a small research project (for an undergraduate or master's student), by being part of a PhD (and possibly part-funded), or by supporting an engineering doctorate (a PhD where the student is sponsored by a company and works for them). To understand what opportunities are available may require the building of relationships. While most companies assume universities require lectures by externals, experience shows what is really needed are projects (research and design), mentors and members of the Industrial Advisory Committee; people who can support and facilitate. By creating a partnership with a university, they may be able to be involved in supporting testing. However, two notes of caution, firstly there is no guarantee that they will be able to do the testing you want, and secondly there is no guarantee of the quality of the output, especially if the project is by a student. Therefore, if the project is urgent, you are much better off working with the university on a commercial basis, but if you are wanting to explore possibilities, ideally that lead to both commercial and academic outcomes, then creating a relationship may be the best way to go (Figure 11.23).

Figure 11.23: SoFSi (Soil-Foundation-Structure-Interaction) laboratory at the University of Bristol which includes a large shake table and a soil structure interaction test pit

y. At a testing house

Around the world there are commercial testing houses, from fire testing to facade testing. If you need a test done and it needs to be done quickly this is probably your best option. It will likely be more expensive than a university, but they will also be more directly set up to carry out your specific test in a timely way (they have no student projects to work around, for example). If you are working with a testing house it is important you get the specification for the work exactly as required, as you would for any other contract. It is also important to work closely with them to define your expectations of who is doing what. Anything you haven't asked them to do, you are doing! Testing houses should also be UKAS accredited[11.17].

z. Existing buildings

Finally, as we will see in Part 4 and as noted in points **e.** and **iv.** you may carry out your tests on an existing building. If the building is in use, then the risks of testing may rise (the consequence of failure may substantially increase) and this should be considered in your risk assessment.

There are also several specialist testing providers who will carry out *in situ* testing, for example, a condition survey of timber. As with all works on an existing building, before you do anything it is important to carry out all necessary risk assessments and, if necessary, building surveys, for example for asbestos (Figure 11.24).

Figure 11.24: Load-testing repaired joists at Dyrham House

11.9 CONCLUSIONS

In this chapter we have looked at a wide variety of considerations of testing. This is to provide a picture of just how much is possible and all the different issues that should be considered. The purpose of the book is to look to the future of design, and while some of this testing is already happening, for many engineering practices this will be unfamiliar territory. However, as noted in the introduction to this part, with the advent of regenerative design and continuous, place-based design, testing is likely to become more prevalent. In Appendix C we have included some information for those considering carrying out a test in the near future.

In Chapter 12 we use the information detailed in Chapters 10 and 11 and reconsider the case study in Chapter 9 of Part 2, as we explore what this could look like for the rammed earth walls in that building.

REFERENCES

11.1 British Standards Institution. *BS EN 1990:2002+A1:2005: Eurocode. Basis of structural design. Annex C and D.* London: BSI, 2005.

11.2 Science and Technology Facilities Council. *Eligibility of Technology Readiness Levels (TRL).* UK Research and Innovation, 2022. Available at: https://www.ukri.org/councils/stfc/guidance-for-applicants/check-if-youre-eligible-for-funding/eligibility-of-technology-readiness-levels-trl/ [Accessed: January 2025].

11.3 ASTM International. *ASTM D2915-17(2022): Standard Practice for Sampling and Data-Analysis for Structural Wood and Wood-based Products.* West Conshohocken: ASTM International, 2022. Available at: https://www.astm.org/d2915-17r22.html [Accessed: January 2025].

11.4 Standards Australia. *AS/NZS 4063.2:2010: Characterization of Structural Timber - Part 2: Determination of Characteristic Values*. Standards Australia, 2010. Available at: https://shop.standards.ie/en-ie/standards/as-nzs-4063-2-2010-129640_saig_as_as_274344/ [Accessed: January 2025].

11.5 British Standards Institution. *BS EN 12512:2001: Timber structures. Test methods. Cyclic Testing of Joints Made with Mechanical Fasteners*. London: BSI, 2002.

11.6 ASTM International. *ASTM D5764-97a: Standard Test Method for Evaluating Dowel-bearing Strength of Wood and Wood-based Products*. West Conshohocken: ASTM International, 1997. Available at: https://www.astm.org/d5764-97a.html [Accessed: January 2025].

11.7 Malkowska, D. *et al.* 'Theoretical and Experimental Study on Laterally Loaded Nailed Bamboo Connection'. *Construction and Building Materials*, 342, August 2022. Available at: https://doi.org/10.1016/j.conbuildmat.2022.127971 [Accessed: January 2025].

11.8 CROSS UK website. Available at: https://www.cross-safety.org/uk [Accessed: January 2025).

11.9 British Standards Institution. *BS EN 1996-1-1:2005+A1:2012: Eurocode 6. Design of masonry structures - General rules for reinforced and unreinforced masonry structures*. London: BSI, 2012.

11.10 Boote, S. and Lynes, A. 'Stone as a structural material. Part 1: mechanical properties'. *The Structural Engineer*, 98(3) March 2020, p6. Available at: https://doi.org/10.56330/VDXN2678 [Accessed: January 2025].

11.11 BRE Trust. *Passivhaus Primer: Airtightness Guide: Airtightness and Air Pressure Testing in Accordance with the Passivhaus Standard*. Watford: Passivhaus, BRE Trust, 2014.

11.12 Blass, H.J. *Timber Engineering STEP 1: Basis of Design, Material Properties, Structural Components and Joints*. Almere: Centrum Hout, 1995.

11.13 The Institution on Structural Engineers. *Manual for the design of timber building structures to Eurocode 5 (Second edition)*. London: IStructE Ltd, 2019.

11.14 Dyson, J. *The Accidental Engineer.* Malmesbury: James Dyson Foundation, 2024. Available at: https://www.jamesdysonfoundation.co.uk/who-we-are/our-story/#:~:text=5%2C127%20PROTOTYPES,world's%20first%20bagless%20vacuum%20cleaner [Accessed: January 2025].
11.15 Halfacree, G. and Everard, B. *Get Started with MicroPython on Raspberry Pi Pico (Second edition).* Raspberry Pi Press, 2024.
11.16 Mola website. Available at: https://molamodel.com/ [Accessed: January 2025].
11.17 UKAS. ISO/IEC 17025 Laboratory Accreditation. Available at: https://www.ukas.com/accreditation/standards/laboratory-accreditation/ [Accessed: January 2025].

CHAPTER 12

Future case study — Part 2

Raffaele De Risi and James Norman

12.1 INTRODUCTION

As we conclude Part 3, we return to the case study from Part 2 (Chapter 9) and consider how we might apply what has been discussed, to the design of the rammed earth walls. This case study is not an attempt to provide a 'how to' but instead to look at how we might take the information and apply it as we move into a regenerative way of working.

While rammed earth is not, in and of itself, regenerative, as we note in the introduction to this part, regenerative design encourages engineers to reimagine design by starting with the question of what is available and what is abundant, which we can use to enable human and living systems to survive, thrive and co-evolve.

Often, the answer to these questions leads to more questions, such as, what if the solution requires a design not covered by our codes of practice? What if we are using timber that hasn't been graded or agricultural waste that is untested? What if we have a new type of concrete which needs classifying?

The following case study outlines the thought processes we might follow.

12.2 THE CASE STUDY

In this chapter, the case study focuses on the rammed earth walls (Figure 9.1). The structure is 11m long, 5m wide and 2.8m tall. The walls are formed out of two C-sections that face each other to create the living space. On the long sides, the walls support the roof, on the short sides they support the overhang and a short back span to provide some additional mass to the wall.

The aim is to design a structure that has as little negative impact on the environment as possible. For this reason, the client has suggested the possibility of using rammed earth. Rammed earth has a very low embodied carbon if the material is site-won. If earth needs to be imported the embodied carbon rises due to transportation emissions. Rammed earth can also, at the end of life, be returned to the ground from where it came and does not become landfill, but only if non-cementitious stabilisers are added, or no stabilisers at all. Therefore, the design is dependent on the material being site-won and of the right soil type to be suitable.

Note: This case study is designed to highlight how one may approach a new or unknown material. It is not designed to be a 'how to' for rammed earth design, and many of the nuances of rammed earth design have not been included in this speculative study.

12.3 DESK STUDY

Before testing and designing the walls, a thorough desk study should be undertaken. This will take two forms. Firstly, research into existing texts on rammed earth design, with a focus on failure types and what needs to be tested (both physically and through the use of analysis models). Secondly, a study into the expected material type to see if it is likely to be suitable — rammed earth is constructed from clay-based site-won material.

12.4 SITE STUDY

Following the initial study, it is agreed that a site investigation will be carried out. This will include testing for rammed earth, as well as allowing for other construction types to be used if the soil is not suitable. You specify the number of trial pits from across the site. On the day of testing, before you take any samples, once the topsoil has been removed, you carry out some initial 'drop tests'. These are qualitative tests to see if the soil is cohesive. Based on your findings, you take a few rubble sacks full of earth back to the lab for further testing.

12.5 INITIAL STUDY

Having taken a number of sacks of site-won material from different trial pits (all carefully labelled) you begin the job of working out if the material will be suitable. Following published guidance, including *Rammed earth: design and construction guidelines*[12.1], you work with a local university to do some initial testing. This includes a particle size distribution, the plastic and liquid limit and measuring the amount of organic matter. The outcome of these tests is favourable and suggests that the soil will be suitable, without any stabiliser, and should not have undue shrinkage. However, at this stage a lot more testing needs to be carried out. To aid in the design, you have also made a few cylinders and measured their density. At the next design team meeting you present the cylinders to the client and their family. One of the main considerations, which you can't control, is the colour of the finished earth. The client is happy with the colour, however, one of their children notices that one of the cylinders smells unpleasant. Rammed earth, typically, contains a small amount of organic matter and if the amount is too high it can lead to an off-putting odour. As the amount of organic matter was below the accepted level, the aroma was not anticipated to be an issue but, in this case, as the child who raised the concern is neurodiverse and is particularly sensitive to smell, the client is keen that this earth is not used or the child may not feel comfortable in the building[12.2]. It is only one of the cylinders which has raised this issue, so they agree that further testing can be carried out (with associated costs) but are keen that organic content and smell are key considerations. After the meeting you reflect on how, as an engineer, you have never had to consider the relationship between neurodiversity and fragrance of a material before, and realise it is an unknown unknown and add it to your material testing schedule as something to consider in the future.

Before carrying out the next round of testing you are keen to do some initial calculations to work out the key parameters. Following design guidance, you discover that the wall could fail in a number of ways, mainly:

During construction — rammed earth walls are most likely to fail after the formwork is removed if the material has not been compacted properly, either due to too much or not enough moisture, so you need the optimum moisture content to achieve maximum density
Under axial compression — for which you will need to establish the compressive strength
Under bending from wind load — assuming the wall has no tensile capacity it will be dependent on its own self-weight to resist overturning so the density will be required for this
Under wearing — rammed earth is susceptible to surface wearing and, although this is a house, so is less susceptible to wearing than other uses (for example, schools) it is still an important consideration

If exposed to wetting — rammed earth is susceptible to erosion if exposed to continued wetting from rain. A variety of strategies can be used, from providing a roof overhang to minimise rain exposure, to covering the earth with a render. In this case the client wants the earth exposed so an overhang for the roof will be utilised.

You plan to make the walls loadbearing, so do some initial design. Based on your reference book a strength of $2N/mm^2$ is assumed with an assumed partial safety factor of six (although this is only for initial sizing purposes; once you have more information you intend to do a reliability analysis).

A very quick analysis suggests that the axial compression will not be an issue. Generally this type of design is governed by the overturning of the wall under wind load.

Based on a number of similar exemplars that you have found through your research you expect the wall width to be 400–500mm. At this stage you decide to go for 450mm.

In addition, you consult a fire engineer to review the risks surrounding the use of rammed earth in the external face. One of the concerns raised was the impact of high pressure water hoses when putting out the fire.

You also consider the long-term loading effects on axial capacity. Based on current research and looking at other properties, you decide to allow for some settlement in the details due to long-term movement, but are less concerned about significant change in performance — although you note more research in this area would be helpful.

12.6 MATERIAL TESTING

Based on the initial study a more in-depth round of testing is to be carried out. These are detailed in Sections 12.6.1–12.6.4.

12.6.1 Optimum moisture content

The optimum moisture content is calculated by using the heavy Proctor compaction test. This method involves compacting five layers of soil by dropping a 4.5kg weight from a set height 27 times on each layer. Each layer is 40mm thick, so the overall height is 200mm. The sample is then dried out and the dry weight is measured. The test is repeated at different moisture contents to establish the maximum dry density (typically we would use 10–14% in increments of 1%). This is a standard geotechnical method for determining the moisture content of a sample which is required to achieve the maximum density sample. The heavy Proctor method provides a lower-bound indication of the maximum dry density (MDD) which is likely to be achieved when using rammed earth. The testing is summarised in Table 12.1 and suggests an optimum dry density of 11%.

Table 12.1: Dry and wet density of the samples following heavy Proctor testing

Moisture content	Dry density (kg/m^3)			Wet density (kg/m^3)		
	Sample 1	Sample 2	Sample 3	Sample 1	Sample 2	Sample 3
7%	1478	N/A	N/A	1581	N/A	N/A
9%	1523	N/A	N/A	1660	N/A	N/A
11%	1564	1596	1583	1736	1772	1757
13%	1498	N/A	N/A	1693	N/A	N/A
15%	1450	N/A	N/A	1668	N/A	N/A

12.6.2 Strength testing

Following the optimum moisture content testing you construct five cylinders, 200mm high by 100mm diameter, using the same compaction method to be used on-site (at this stage you are not sure so you opt for manual ramming on the basis that it won't compact as well as mechanical ramming). As the earth will be prepared on-site and you are concerned about the impact of moisture content you decide to make the cylinders at different moisture contents, −4%, −2%, 0%, +2%, +4% of optimum moisture content determined by the heavy Proctor test. You also decide to make two additional cylinders at the optimum moisture content. The strength test results for the cylinders are provided in Table 12.2.

Table 12.2: Cylinder strengths at different water content

Water content (difference from optimum)	Cylinder strength (N/mm^2)		
	Sample 1	Sample 2	Sample 3
7% (−4%)	1.4	N/A	N/A
9% (−2%)	1.9	N/A	N/A
11% (0%)	2.4	2.0	2.2
13% (+2%)	2.5	N/A	N/A
15% (+4%)	1.8	N/A	N/A

12.6.3 Wearing surface tests

Alongside the strength testing, you also carry out abrasion testing in accordance with the methodology presented in *Rammed earth: design and construction guidelines*[12.1]. The results suggest that the surface will be adequate, but over time any sealant that has been applied to the exposed face of the wall (typically beeswax) will need to be reapplied and that, in some circumstances, the wall may require site-won clay to be applied to fill any visible signs of wearing. You communicate this to the client, as this is not typically the case.

12.6.4 Organic matter and aroma testing

Based on your findings from the client meeting, you carry out some further research into organic matter and aroma. You decide to follow a testing protocol to calculate the level of organic matter from different site locations to work out where you can and can't extract soil, but you also invite the client's family to do a blind smell test of the different samples to see if any of the samples have an unpleasant aroma. The results of both tests confirm that one location in particular is not suitable for extraction.

12.7 DESIGN

12.7.1 Designing for axial load

Having completed the compression testing of the samples, you can now design for axial load. Taking the data from Table 12.2 you decide to analyse just the three values for the optimum moisture content as these are the only repeated tests. You reflect on whether you should have either collected more soil at the time or carried out a secondary investigation, but both approaches have their downsides. Collecting more earth at the point of the investigation requires the earth to be stored and then safely disposed of while you run the initial tests to see if the earth is suitable. Doing a second investigation creates more disruption on the site and an additional fee. While there is no right answer, on balance you are happy you took an acceptable amount. You analyse the data and find a mean of $2.2N/mm^2$ and a standard deviation of $0.2N/mm^2$. Based on the initial design and calculated values, it appears compression will not govern design, which you verify through analysis.

Note: As this situation is not the governing case the reliability analysis provides no further insights so we have excluded it from the example. In a real design you would need to explicitly verify this.

12.7.2 Designing for lateral load

The lateral load on the wall leads to the wall overturning. Rammed earth has little to no tensile capacity (it is so low it is very hard to measure experimentally), so when tension occurs in the wall, a crack forms, which, as designers, we have designated the point of failure (the wall may have some post-failure strength but further testing would be required to prove this). There are many analysis techniques available, but for simplicity, the wall is treated as a propped cantilever (the roof acting as a diaphragm and the perpendicular walls taking the load to the ground). Using a simplistic Euler-Bernoulli analysis[12.3], the bending in the wall from the lateral load is calculated to give a stress, which is resisted by the gravity load on the wall (Figure 12.1).

Figure 12.1: The simplified model of the wall and equations for maximum and minimum stress

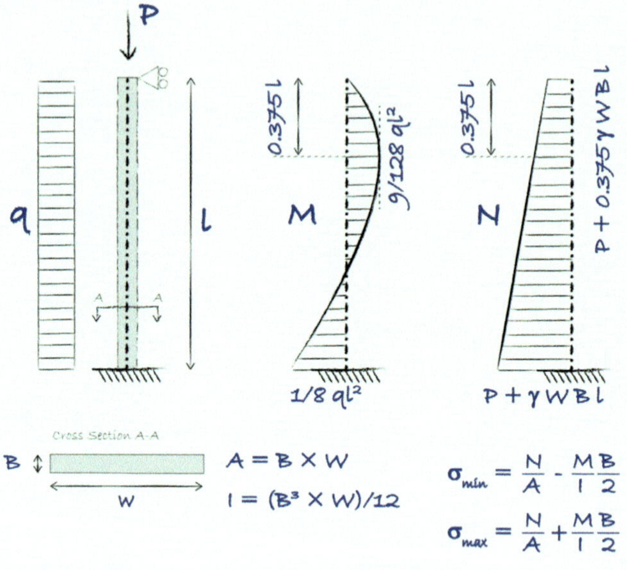

Although the wall is a three-dimensional structure, it is also simplified, and a 1m strip of the wall is considered to span in one direction. Even with all this simplification there is still much uncertainty to consider. The following parameters were modelled:

The load from the roof, P. This is consistent with the load presented in Section 9.2.6.

The lateral load, q, is representative of the wind pressure. The wind pressure is modelled as:

$$q = 0.5 \times air\_density \times wind\_velocity^2 \qquad 12.1$$

the wind velocity is described as a 'Weibull distribution' with scale and shape factor of 15 and 1.5, respectively.

The wall thickness, B, is a design parameter.

Finally, you look at the density of the wall. This is complex as rammed earth density varies with time as it absorbs and releases moisture. However, for simplicity, the wet weight of the wall was calculated with a mean of 1,755kg/m^3 and a standard deviation of 18.1kg/m^3.

A Monte Carlo method analysis is then run for different values of wall thickness propagating all the described uncertainties. The outcome of Monte Carlo simulations for wind velocity, rammed earth self-weight, and the corresponding safety margin for a wall of 300mm thickness at the base is shown in Figure 12.2.

Figure 12.2: Monte Carlo method simulations for wind velocity and self-weight and safety margin for a wall 300mm thick at the base for tensile limit state

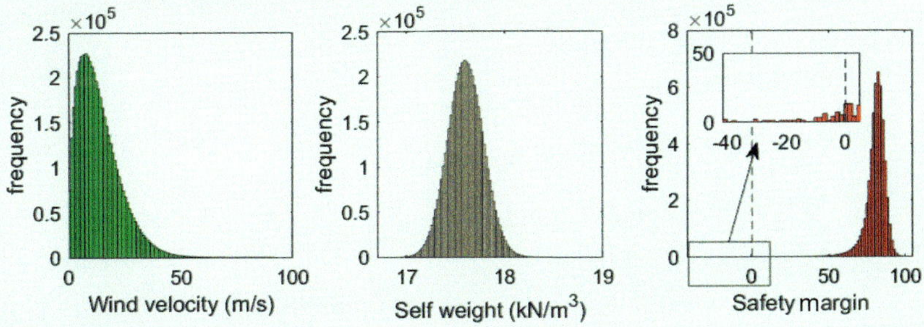

The Monte Carlo method analysis was repeated for several different wall thicknesses. Failure (the wall going into tension) was considered at both the base of the wall (bottom) and at the 3/8 height position (top). These represent the positions where the moment is at a local maximum/minimum. As can be seen, failure is far more likely at 3/8 from the top. This is due to lack of self-weight of the wall (Figure 12.3). In theory, an ideal wall thickness would be approximately 420mm as it achieves the target reliability index of 3.8 over a 50-year reference period. However, there are other considerations, including the size of a standard wacker plate and digger bucket. As a result, a 450mm thickness wall is specified; this provides a tolerance of 30mm from the minimum thickness, which achieves the target reliability. It is noted that 450mm is relatively narrow physically to climb inside the formwork and ram the earth, but based on a review of similar projects it is agreed that it is achievable.

Figure 12.3: Reliability analysis of walls of various thickness to find the optimum wall thickness to achieve an acceptable risk of failure

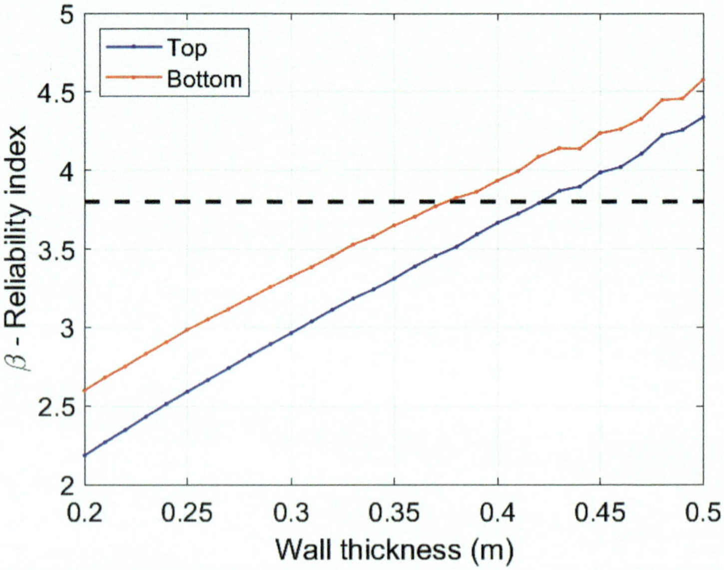

12.8 ON-SITE TESTING

Finally, once the contractor is appointed you ask them, as part of their initial works, to make some trial panels. These will be used in part for quality control purposes, and to ensure the client is happy with the appearance and finish.

In addition, as the material is quite susceptible to effects of moisture content, especially during construction, you request three random cylinder tests during the build to ensure strength and moisture content are acceptable, which, following lab testing, they are.

12.9 CONCLUSIONS

This case study, which focuses on a simple wall design, hopefully highlights several considerations when using a non-codified material. From the necessity of carrying out a desktop study and doing some initial investigative studies to performing a suite of tests to establish key properties, it highlights how we need to stay open to different failure types, and not assume that the material will behave in a certain way. It demonstrates how a probability of failure can be calculated for a specific design case and how we can run multiple Monte Carlo method analyses to find the optimum design.

However, it also highlights that while reliability analysis is a key component in design, the engineer needs to do much more than simply carry out some calculations; utilising a wide set of skills, from exploring diverse material properties and considering construction requirements, to communicating openly with the client and involving them in the decision-making process.

REFERENCES

12.1 Walker, P. *et al. Rammed Earth: Design and Construction Guidelines*. Bracknell: IHS BRE Press, 2005.

12.2 The Institution of Structural Engineers. *Inclusive design for structural engineers*. London IStructE Ltd, 2024.

12.3 Öchsner, A. *Classical Beam Theories of Structural Mechanics*. Cham: Springer Nature, Switzerland AG, 2021.

Part 3 Conclusion — Classifying and testing materials

In this part, we have discussed, in detail, the characterisation of materials and the use of testing to facilitate this. In Chapter 10, we explored the characterisation of materials, provided a long (but by no means exhaustive) list of questions and, where necessary, provided a variety of different probability distributions and tests to understand which is the most applicable.

In Chapter 11 we turned our attention to testing, and covered a very wide set of considerations for testing, from how to find information on the materials we are interested in, to different types of tests we may want to consider utilising. While testing codes and procedures exist for specific materials, and testing methods are provided for new products which are going to market (which is a very different case from exploring a new material or combination on a project), a simple, unified approach does not exist for projects, and probably never will as the depth and breadth of the questions the engineer needs to ask and test for is highly dependent on the material itself. However, we would highlight the pitfalls of assuming we know what we need to test for based on prior information — especially the assumption that the only information we need is the stress-strain relationship based on our current use of a limited pallet of materials. We would also highlight that we need to test to work out the questions we need to ask before we then test to answer them. This double approach can take longer and be frustrating but will also reduce the chance of an unknown cause of failure appearing later in the design process.

While regenerative design may lead to a much more diverse palette of materials, it also calls for us to use the rich resources we have in the existing buildings we already own. In Part 4, we explore how we think about uncertainty within reuse.

SECTION 15 : 3
THANK YOU
98.9%
WASTE RECYCLED
YOU CAN'T THROW SOMETHING AWAY – THERE IS NO "AWAY"
CCS CARD
NEW START SCHEME
6.03.47
DAYS SITE
TENDANT:
HARITY
DELEYE
LAST ACCIDENT
JAN 23rd 2035

Part 4
Reuse

Our industry is changing. Reuse of existing buildings and building components is moving from a niche that is predominantly covered by conservation specialists, to becoming a core discipline of structural engineering practices. This change has been captured in the IStructE's book *Circular economy and reuse: guidance for designers*[P4.1].

Engineers are aware of the uncertainty and risk inherent in an existing building and building components, from degradation of materials to poor quality of construction. However, they may be less aware of the risks which face a new-build, which now the building is built, can, at least in theory, be measured and quantified, from out-of-plumb and tolerance of construction to material properties. How do we deal with this shift in uncertainty? How do we account for the unknowns that are now known and the new uncertainties that existing buildings present?

In Part 4 we explore the uncertainty that engineers face in reuse, how to reduce and manage this uncertainty, and present another statistical approach, Bayesian updating, which may be helpful in this growing field. While this part focuses on buildings, similar methodologies could also be applied to building components.

REFERENCE

P4.1 The Institution of Structural Engineers. *Circular economy and reuse: guidance for designers*. London: IStructE Ltd, 2023.

CHAPTER 13

Assessment and reuse of existing buildings

Flavia De Luca and Gavin Knowles

13.1 INTRODUCTION

This chapter aims to review the concepts introduced in the previous parts of this book to relate to existing buildings and reuse. While the methodological aspects are not affected by moving from a new construction to an existing building, the level of confidence regarding knowledge of the structure introduces a new source of uncertainty.

The chapter is also intended to provide a collection of relevant international codes and guidance for the specific case of existing buildings, with a focus on resources available for the UK and specific reference to *ISO 13822:2010 — Assessment of existing structures*[13.1]. The very sparse nature of information on existing buildings typically leads to the lack of very quantitative prescriptions in codes and standards, and to the extensive use of design flowcharts and qualitative descriptions of the steps to be considered when dealing with an existing structure.

13.2 THE CASE FOR REUSE

All buildings need maintenance; some require more than just this and are refurbished beyond their original use. Consider the building shown in Figure 13.1 which used to be part of a large area with derelict Grade II listed buildings used up to 1989 to house the Football Host Company and since then abandoned (Figure 13.1a). It has now been repurposed into a residential development integrated within the local community (Figure 13.1b).

Figure 13.1: Example of refurbishment and change of use

(a) (b)

With any existing building stock, the risk to its longevity is not just its structural adequacy or maintenance and upkeep but also how it is used, whether it brings in a rental income or how it may be conserved for future generations. Change of use of an existing building can have a significant impact beyond its own physical space, affecting the surrounding area and having both positive and negative consequences (e.g., requalification, gentrification, change in real estate market of the surroundings).

Existing buildings can be amazing; they can be grand and complicated or simple and understated and can continue to inspire if we look after them. There are potentially more challenges for the engineer working on an existing building, depending on its state of repair or its intention to become something different and serve a different purpose. We must fully get to know our building — in structural terms this means assessing condition and capacity. This chapter helps the reader to appreciate the challenges embedded into approaching existing buildings and will also provide some tools to qualify and quantify some inherent uncertainties.

Before getting into some of the technical tools that could help engineers in dealing with known and unknown uncertainties associated with existing buildings, their assessment and retrofitting or refurbishment, the first question is: "Why do we want to keep existing buildings/infrastructure?" Buildings pose a valuable investment to the owner and can have value far beyond their financial return. Many existing buildings have conservation value to be retained for future generations to enjoy.

Note: You can find out the listing status of any building in England at the Historic England website. Similar information exists for Wales, Scotland and Northern Ireland[13.2].

Some buildings have higher conservation value based on how they are built and how rare they are. Existing buildings are graded into groups depending on their perceived historic value. National Heritage List for

England (NHLE) by Historic England provides information on listed buildings and grades.

We should also consider saving or repurposing an existing building which has sustainable value. One building reused saves a new building being built. From a client's viewpoint it might also allow operational continuity and release funds for use elsewhere in their business. The ideology of using what we have aids the movement towards net zero[13.3].

The aim of this chapter is to guide the structural engineer through the process of working with an existing building and manage its risk. We will highlight some common areas of risk and the approach to reading an existing building; we will look at how understanding (as far as possible) an existing building's form and function can reduce risk not just for the engineer but also for the client. This includes consulting historic codes which align with the age of the building. Importantly, early decisions on the building's future purpose will also be addressed and why these are important in futureproofing the building and reducing structural and financial risk.

Working on existing buildings can be split into three categories:

- Assessment (no physical works)
- Modifications
- Change of use

Existing buildings also pose a specific challenge regarding reliability. The general principles discussed in previous parts of this book can be applied to existing buildings, but with some additional and different uncertainties that require alternative methodologies and/or different performance levels.

13.3 WHAT ARE THE RISKS WITH EXISTING BUILDINGS?

Uncertainty and risk are embedded in all construction work. With new buildings, you can control the materials you are designing with. However, existing buildings have the added risk of finding things on-site which are unexpected and you may have to deal with poorly-maintained materials. Well-maintained buildings are not risky if they have been carefully designed and constructed. There are a few reoccurring issues around understanding existing buildings which are helpful to highlight.

Changes in structural understanding or codes can highlight previously undiagnosed risks. This can happen as a result of a specific event which provided disastrous consequences. An example in recent UK history was the Grenfell Tower fire in 2017[13.4] which led to a change of Building (Amendment) Regulations in 2018. The Grenfell Tower fire is one of the UK's worst modern disasters and led to significant consequences for all high-rise buildings in the UK. New regulations and codes are often the result of lessons learnt from disasters or simply from experience collected over time.

Existing buildings have often been designed according to obsolete standards and the first step is to appraise their performance with respect to more recent standards, considering any effect that time and external actions have had on them. Appraisal is 'checking the adequacy of an existing structure', which sometimes becomes necessary for different reasons, such as change of use, structural alterations etc.[13.5].

As a result of a detailed appraisal the client can decide the most appropriate course of action: refurbish, demolish and rebuild, or do nothing. The nature of refurbishment is identified from understanding the life of the building and how well-maintained it is. Figure 13.2 highlights how the structural performance of a building is related to its longevity (design life). As time passes, the building asset deteriorates and if no intervention is carried out in the form of routine maintenance, the building will not meet its performance criteria. To reduce risk to the building a scheduled repair and maintenance scheme should be in operation, so that deterioration of the building is managed and the risk to the client of losing their asset is reduced. Figure 13.2 also highlights the ability of the building to be restored or overhauled and enhanced to increase its performance, which could extend the life of the building and therefore reduce the risk of losing the asset.

Figure 13.2: Asset performance over design life

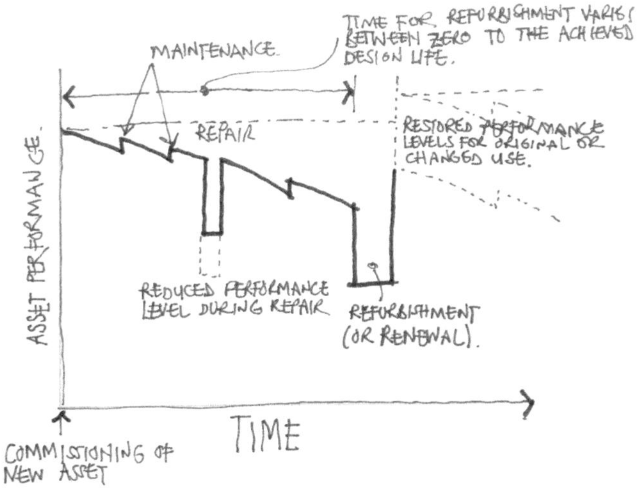

Existing buildings can be risky as we can't always see the full structure. A comprehensive investigation of the structural form and materials is necessary to understand the existing building you are working with. While geometry will need a full definition for structural details (e.g., reinforcement) and material properties, specific quantitative thresholds can be used to achieve the required knowledge level (e.g., CEN 2005[13.6]). This allows us to make considered and educated decisions on how the loads are transferred through the building and how large these loads may be. Later in this chapter we will discuss approaches to discovering as much information as we can about the building.

There may be issues with deterioration that could be detrimental to structural adequacy. A complete review of the structure is fundamental to highlight such areas. A good knowledge of the building is key here. This can be achieved through a site investigation, measuring dimensions of structural elements and testing of material properties (Chapter 11).

Other areas of uncertainty and risk can come from the building's lack of relevance to new codes and methods of construction. The methods of original construction may constrain the ability of the refurbishment to be carried out in a safe manner.

Other challenges may be short timelines (programmes) to complete the refurbishment works. The issue here is that unexpected problems may halt or delay the works. To manage this risk, more extensive and potentially expensive investigation works should be prioritised early in the project. Connected with these issues is the management of the client's perception of why the project is risky and how this translates to risk on-site, in terms of tackling delays.

Table 13.1 explores change in use; one of the man reasons for assessment and intervention on an existing building. There are other reasons for assessment including: extension of design working life of the building (a reliability check required by authorities or insurance companies) and in cases of structural deterioration due to time-dependent actions (e.g., corrosion) or structural damage due to accidental actions. This is the remit covered by the codes discussed in the following sections.

Table 13.1: Risk classification table for existing buildings in case of reuse and their risk profile

Building type	Original use	Typical conversion	Challenges	Risk profile/ risk owner
Domestic house (all periods)	Housing	Rear extension Loft extension	Not many high-risk tasks to conversion or extension. The key challenge is working with a competent design team and contractor. Temporary works are an important part of these works which are often overlooked	Low/owner
Industrial (Victorian or modern)	Factories Process plants Warehouses	Residential Offices Museums Community spaces	Due to age, there is often limited knowledge of the structure. There may be a lot of deterioration to the construction materials, hazardous materials and material variability. However, rectifying issues is often simple engineered repairs based on a limited palette of materials	Medium–high/ client and contractor
Church	Religious	Residential Community space Religious	Careful consideration of the historic fabric may prevent major interventions	Medium/owner

(Continued)

Table 13.1: *Continued*

Building type	Original use	Typical conversion	Challenges	Risk profile/ risk owner
Commercial	Offices Retail (1950s–present) Warehouses	Refurbishment Vertical extension	Time is of the essence to refurbish this building type. Tenants may be present for the works and the refurbishment may need to be phased. Often code provision is limited	High/funders
Education	Schools Universities	Extensions Refurbishment Reconfiguration	Quality is important and reconstruction timing is key to suit education holiday periods. Phasing is often necessary. Sometimes prefabricated elements can be used	Medium/owner
Critical	Hospitals Police stations Bridges	Extension Maintenance	These types of infrastructure have longer life spans and need to be maintained, requiring special consideration, and are rated separately in the Eurocodes	High/owner

One final note is that on the assessment and rehabilitation of existing buildings a significant lesson learnt comes from the specific case of bridges and heritage structures — on which engineers over the years have had to develop specific maintenance and inspection techniques and very demanding requirements to protect the integrity of the original structure. These two main aspects, in conjunction with the necessity to assess performance of structures against changed loads (e.g., earthquake and traffic) have all contributed to the development of bespoke approaches for existing buildings, accommodating their peculiarities beyond the conventional challenges faced when dealing with new constructions.

13.4 CODES AND RECOMMENDATIONS FOR EXISTING BUILDINGS

Assessment of existing buildings has become more prominent in the last 20 years, and it is going to become even more important in the context of more sustainable approaches in the construction industry and control of carbon emissions.

In terms of codes, recommendations and regulations, each country has, over the years, developed its own specific set of recommendations. One of the earliest and most complete reference documents is represented by the RILEM Publication by the Joint Committee on Structural Safety (JCSS) dating back to 2001[13.7]. This document was put together within the overall effort of the JCSS and addressed very specific questions for existing structures, beyond the specificity of buildings, such as: "What type of inspections are necessary?", "What analyses shall be performed?", "What are the risks involved in further using the structure?" and "What are the risk acceptance criteria to be considered?" as discussed in *Assessment of existing structures — on the applicability of the JCSS recommendations*[13.8].

In terms of international standards, ISO 13822:2010[13.1] represents the main reference addressing the issue of existing structures in the modern context of economic, social and sustainability considerations, establishing the language and overall principles for the assessment of existing structures within a reliability context. Annex A of that document also establishes a 'hierarchy of terms' to be used and encompasses the different aspects of assessment and rehabilitation of structures (Figure 13.3).

Figure 13.3: Hierarchy of terms

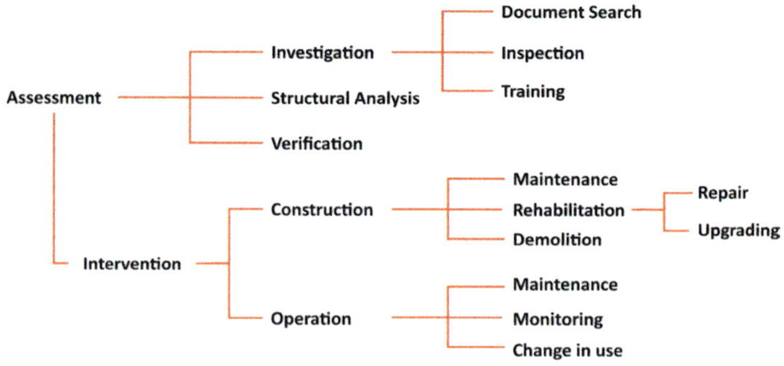

Derived/adapted from Annex A of ISO 13822.

This hierarchy provides a preliminary overview of the different steps for general assessment of existing structures, covered in detail in the flowchart presented in Annex B of ISO 13822 (Figure 13.4 of this book). The assessment of existing structures has a number of additional nested iterations with respect to the conventional, iterative procedure used for design and verification of new constructions. In particular, the objectives and scenarios (i.e., change in structural conditions or actions) are defined in a bespoke way given the specific project, and the way in which preliminary assessment is performed. There are at least three iterative steps related to whether a detailed assessment is necessary, whether further inspection is required and finally whether a target reliability is reached, and interventions need to be implemented. Figure 13.4 provides an overview of the possible intervention options that an engineer has when dealing with an existing structure.

The use of flowcharts is common in documents covering existing structures, given the multiple situations that can be encountered in the assessment and intervention processes e.g. Section 4 of *Appraisal of existing structures*[13.5].

Figure 13.4: Flowchart for the general assessment of existing structures

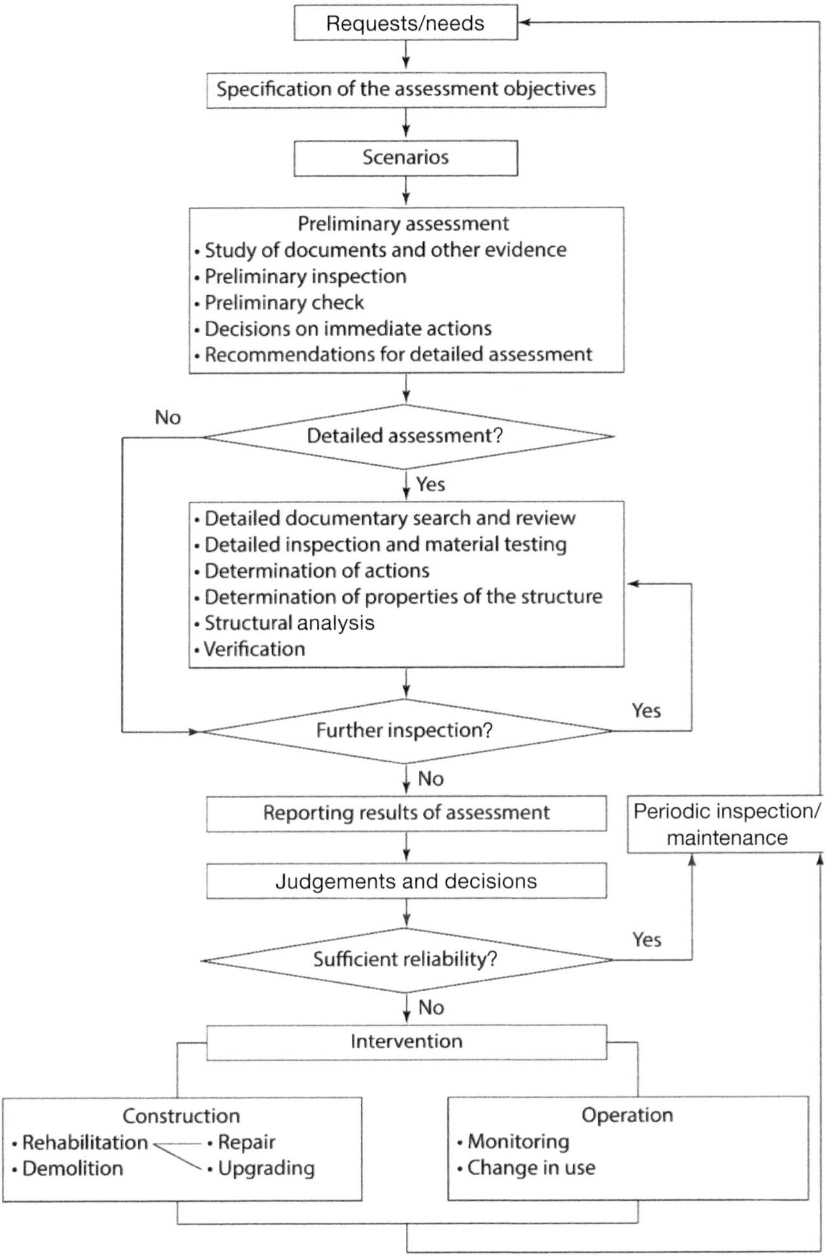

Derived/adapted from Annex B of ISO 13822.

For the first generation of Eurocodes, assessment of existing buildings appears explicitly in Eurocode 8, Part 3[13.6] where the analytical seismic assessment of buildings is covered in a normative way with an approach that moves from the partial factors and relies on the concept of 'knowledge level', building upon different studies, such as *Assessing existing buildings with Eurocode 8 part 3: a discussion with some proposals*[13.9], but also embedding some of the JCSS work and earlier versions of ISO 13822 documentation made on assessment of existing buildings.

Within the activities of the European Technical Committee 250 (CEN/TC250) — the Eurocode structural committee covering 'the standardisation of structural and geotechnical design rules for building and civil engineering works taking into account the relationship between design rules and the assumptions to be made for materials, execution and control', working group 2 has covered the assessment and retrofitting of existing structures, releasing a policy report in 2015, *New European technical rules for the assessment and retrofitting of existing structures*[13.10] which led to the release of the technical specification document in July 2020, *CEN/TS 17440 Assessment and retrofitting of existing structures*[13.11]. This document is developed within the overall context of Eurocodes and relies significantly on concepts and definitions presented in ISO 13822[13.1] and the JCSS document on assessment of existing structures[13.7].

The aim of CEN/TS 17440 is to cover existing structures (all types of buildings and construction works, including geotechnical aspects) and structures exposed to various actions, identifying the specific aspects set up by Eurocodes, and mainly developed for new construction that can be adapted to existing structures, identifying and distinguishing the differences of these two contexts.

As for all documents developed under the aegis of the CEN/TC 250, the document embeds rules and principles developed in different countries leaving specific mandate to develop National Annexes. For the aspect of existing structures, the policy report[13.10] covered an overview of documents and regulations employed in basic practice for different countries. This is a very important effort since it also emphasises where, traditionally, in each country the specific problem of existing structures has been covered and through which type of documentation. For specific structural materials, like reinforced concrete structures, the main Eurocodes have annexes covering how to adjust rules and prescriptions to the case of existing buildings. One example is Annex I of Eurocode 2 where supplementary guidance for the assessment of reinforced concrete structures is provided, BS EN 1992-1-1:2023 - Eurocode 2[13.12].

In the UK, as summarised in the country table of the 2015 policy report, there are no specific national 'building' regulations concerned solely with the assessment and retrofitting of existing structures, but some requirements relating to existing buildings are set down in the Building Regulations. The policy report[13.10] provides an overview of the regulatory requirements and associated professional technical guidance which has been developed for the assessment of existing buildings and structures. For the UK, there is established national guidance concerned with assessment and retrofitting of some specific classes of existing structures, such as highway bridges. Table 13.2 provides a summary of documentation available for buildings as discussed in the policy document.

Table 13.2: Summary of existing UK regulations and standards for assessment of buildings

Regulatory requirements	Technical guidance on structural appraisal
1 The current legislative requirements for the structural performance of buildings are established in Part A (Structure) of Schedule 1 of The Building Regulations 2010 for England and Wales[13.13]. There are separate (but generally similar) legislative requirements for Scotland and Northern Ireland 2 Guidance on meeting the statutory requirements established in Part A of The Building Regulations is given in Approved Document A (currently the 2004 edition; incorporating the 2004, 2010 and 2013 amendments), which establishes the need for certain uses of existing buildings to comply with the requirements of Part A (Structure) of Schedule 1 of The Building Regulations 3 Large panel system-built (LPS) dwelling (housing) blocks have been treated as a special class of building in the UK since the partial collapse of Ronan Point in 1968. The requirements for undertaking their structural assessment are set down in the then Ministry of Housing and Local Government Circulars (MHLG) 62/68 and 71/68 (1968)[13.14] 4 Other than these there are no explicit requirements defined for the structural assessment of existing buildings apart from those that arise in relation to health and safety-related legislation (primarily The Health and Safety at Work Act (1974) and subsequent Acts)[13.15]	1 Section 1 of Approved Document A indicates that in these circumstances appropriate technical guidance on structural appraisal is given in: a. BRE Digest 366: Parts 1 to 4: Structural Appraisal of Existing Buildings (2012)[13.16], and b. The Institution of Structural Engineers' publication *Appraisal of Existing Structures* (2010)[13.5] These documents define procedures that may be used to undertake an assessment of an existing structure and some aspects of the calculation procedures that might be employed. 2 BRE Report 511: *Handbook for the Structural Assessment of Large Panel System (LPS) Dwelling Blocks for Accidental Loading*[13.17] defines revised performance requirements (updating those given in MHLG Circulars 62/68 and 71/68) and provides technical guidance on structural appraisal procedures that may be employed

The main topics to be covered for the assessment of existing structures are:

- Methodology of collecting, evaluating and updating data
- Recommendations for application of partial factor method and direct use of probabilistic methods consistent with those for new structures
- Target reliability level of existing structures taking into account residual working lifetime, consequences and costs of safety measures
- Assessment based on satisfactory past performance
- Resistance models for the assessment of structural elements constructed with different material, detailing provisions and tolerance from those assumed in design standards

These aspects are covered in CEN/TS 17440[13.11] and valuable applicative examples are provided in the underpinning documents such as JCSS[13.7] on the assessment of existing structures. In particular, Annex C of CEN/TS 17440 provides an overview of target reliability and partial factors as complementary guidance to the aspect of reliability management (Section 4.1) that also specifies that the target value of reliability index (β_t) related to a reference period can be set by the National Annex for use in a country.

Given the relatively recent release of the CEN/TS 17440, an engineer interested in applying the 'assessment value method', 'probabilistic methods' or the 'risk assessment method' as discussed in CEN/TS 17440 can refer to values and applicative examples in ISO 13822 and JCSS. In particular, Annex F of ISO 13822 provides an illustration of target reliability levels for existing structures (Table 13.3).

Table 13.3: Illustrations of target reliability levels

Limit states	Target reliability index, β	Referenced period
Serviceability		
Reversible	0.0	Intended remaining working life
Irreversible	1.5	Intended remaining working life
Fatigue		
Inspectable	2.3	Intended remaining working life
Non inspectable	3.2	Intended remaining working life
Ultimate		
Very low consequence of failure	2.3	L_S years*
Low consequence of failure	3.1	L_S years*
Medium consequence of failure	3.8	L_S years*
High consequence of failure	4.3	L_S years*

* L_S is a minimum standard period for safety (e.g., 50 years)

CEN/TS 17440 also covers interventions in Section 11 using the hierarchy of terms provided in Annex A of ISO 13822 (Figure 13.3). The aspects covered are very general, applicable to all kinds of existing structures and reliant on basic principles for retrofitting and/or rehabilitation as discussed in the policy document[13.10]:

- Compatibility of structural components and materials of the old structure and the new structure in view of structural behaviour and durability, which leads to optimisation with synergic action
- As far as historic structures under protection of monuments are concerned: reversibility of the structural intervention so that future generations can modify the technology without any damage to the features of the historic construction

Specific assessment and retrofitting methods are linked with regional peculiarities reflecting shortcomings in old national design or execution standards for the use of materials or building techniques.

The aspect of old national design or execution standards is very important for existing buildings where knowledge of the timeline of country regulations and practices becomes a fundamental aspect for assessment and rehabilitation of existing buildings.

While it is reassuring for engineers to know that there are prescribed levels of reliability and codes that address the approaches required to assess existing buildings, we should not rely on these exclusively, weighing these requirements with engineering judgement. Ultimately as engineers, we will need to decide if something is safe. The available codes should be used to prove this to us once we have reasonable confidence this is the case, rather than be seen as a replacement for our own judgement.

13.5 CONCLUSIONS

In this chapter we have explored why reuse is important, why the uncertainties differ from those considered in new buildings and what codified frameworks exist for assessing existing buildings. We have also highlighted the inherent complexity in assessing existing buildings and why a simple unified approach is not possible. In Chapter 14 we will explore how we gather the necessary data to start to understand the design intent, the constructed building (and how it deviates from the design), the condition of the materials and the uncertainty all this leads to.

13.6 FURTHER READING

As discussed previously, the topic of assessment and rehabilitation of existing buildings can be investigated from different perspectives and one of these is also the historical appraisal of regional practices. While it is not possible to provide a comprehensive list of further reading on the topic, the following is a list of additional documents, seminal papers and books that were not referenced in the body of the chapter, and are provided to expand the topic of the appraisal of existing buildings in the UK.

- Allen, D.E. 'Limit states criteria for structural evaluation of existing buildings'. *Canadian Journal of Civil Engineering*, 18(6), December 1991, pp995–100. Available at: https://doi.org/10.1139/l91-122 [Accessed: January 2025].
- Beckmann, P. and Bowles, R. *Structural aspects of building conservation (Second edition).* Abingdon: Routledge, 2024.
- CIRIA. *Structural renovation of traditional buildings (CIRIA Report 111).* Amended ed. London: CIRIA, 1994.
- Carriageworks Action Group website — case study. Available at: https://carriageworks.org.uk/ [Accessed: January 2025].
- Ellingwood, B.R. 'Reliability-based condition assessment and LRFD for existing structures'. *Structural Safety*, 18(2–3), 1996, pp67–80. Available at: https://doi.org/10.1016/0167-4730(96)00006-9 [Accessed: January 2025].
- Fox, M. Kam, W-Y. and Grant, D. *Seismic Retrofit of Existing Buildings*. Leeds: Emerald Publishing Ltd, 2024.
- Historic England. *Practical Building Conservation — Timber.* Abingdon: Rouledge, 2012.
- Historic England. *Practical Building Conservation — Stone*. Abingdon: Rouledge, 2012.
- Historic England. *Practical Building Conservation — Metals*. Abingdon: Rouledge, 2012.

- Historic England. *Practical Building Conservation — Concrete*. Abingdon: Rouledge, 2012.
- *fib. Partial factor methods for existing concrete structures (fib Bulletin 80)*. Lausanne: *fib*, 2016. Available at: https://www.fib-international.org/publications/fib-bulletins/partial-factor-methods-for-existing-concrete-structures-161-detail.html [Accessed: January 2025].
- Foster, J. 'What can you do if you are convinced a structure will work but can't prove it to code'. *The Structural Engineer*, 99(6), June 2021, pp18–22. Available at: https://doi.org/10.56330/VWVB2904 [Accessed: January 2025].
- Mathew, M. 'Analysing existing structures: a brief introduction'. *The Structural Engineer*, 99(6), June 2021, pp14–17. Available at: https://doi.org/10.56330/XDWP5331 [Accessed: January 2025].

REFERENCES

13.1 *ISO 13822:2010: Bases for design of structures. Assessment of existing structures*. Geneva: ISO, 2010.

13.2 Historic England website for listed buildings. Available at: https://historicengland.org.uk/listing/the-list/ [Accessed: January 2025].

13.3 HM Government. *Net Zero Strategy: Build Back Greener*. Available at: https://www.gov.uk/government/publications/net-zero-strategy [Accessed: January 2025].

13.4 BBC News. *Grenfell Tower: What Happened*. Available at: https://www.bbc.co.uk/news/uk-40301289 [Accessed: January 2025].

13.5 The Institution of Structural Engineers. *Appraisal of existing structures (Third edition)*. London: IStructE Ltd, 2010.

13.6 *CEN (2005) EN 1998-3-1-4: Eurocode 8, Design of Structures for Earthquake Resistance, Part 3, Assessment and Retrofitting of Buildings*. Brussels: CEN, 1998.

13.7 Joint Committee on Structural Safety. *Probabilistic Model Code*. JCSS, 2001. Available at: https://www.jcss-lc.org/jcss-probabilistic-model-code/ [Accessed: January 2025].
13.8 Diamantidis, D. *Assessment of Existing Structures - On the Applicability of the JCSS Recommendations.* Conference: Aspects of Structural Reliability, January 2007.
13.9 Pinto, P.E. and Franchin, P. *Assessing Existing Buildings with Eurocode 8, Part 3: a Discussion with Some Proposals.* January 2008. Available at: https://www.researchgate.net/publication/239777109_Assessing_existing_buildings_with_Eurocode_8_Part_3_a_discussion_with_some_proposals [Accessed: January 2025].
13.10 Luechinger P. *et al. New European Technical Rules for the Assessment and Retrofitting of Existing Structures.* January 2015. Available at: https://doi.org/10.2788/052881 [Accessed: January 2025].
13.11 *CEN/TS 17440:2020: Assessment and Retrofitting of Existing Structures*. Brussels: CEN, 2020.
13.12 British Standards Institution. *BS EN 1992-1-1:2023 - Eurocode 2. Design of concrete structures. General rules and rules for buildings, bridges and civil engineering structures.* London: BSI, 2023.
13.13 HM Government. *The Building Regulations 2010. Approved Document A: Structure*. Available at: https://www.gov.uk/government/publications/structure-approved-document-a [Accessed: January 2025]
13.14 Soane, A. 'Structural-Safety.org — Once is happenstance'. *The Structural Engineer*, 90(8), August 2012, p1. Available at: https://www.istructe.org/journal/volumes/volume-90-(2012)/issue-8/professional-guidance-structural-safety/ [Accessed: January 2025].
13.15 HM Government. *Health and Safety at Work etc. Act 1974.* Available at: https://www.legislation.gov.uk/ukpga/1974/37 [Accessed: January 2025].
13.16 Matthews, S. *BRE Digest 366: Structural Appraisal of Existing Buildings Including for a Material Change of Use.* Bracknell: IHS BRE Press, 2012.
13.17 Matthews, S. and Reeves, B. *Handbook for The Structural Assessment of Large Panel System (LPS) Dwelling Blocks for Accidental Loading*. Bracknell: IHS BRE Press, 2012.

CHAPTER 14

Gathering data

Gavin Knowles

14.1 INTRODUCTION

Collection of data and information to acquire the right level of knowledge on an existing building is fundamental. This also represents a cost in terms of time and resources for engineers, but it provides significant added value for reduction of uncertainty on the *status quo* of the building. The way we hold/archive data can make it easier for adaptive reuse going forward.

14.2 FINDING OUT ABOUT THE BUILDING – TIPS AND PRACTICES

An initial way to reduce risk when redeveloping existing buildings is to find as much information as possible on the building. Key elements to consider are what the historical use and current use of the building is, and whether original and subsequent drawings are available. This data collection exercise is also aided by speaking to the right people. This is covered in the 'investigation' — the first branch of the hierarchy of terms provided in Figure 13.3 and defined in ISO 13822[14.1] as 'collection and evaluation of information through inspection, document search, load testing and other testing'. The first step is a document search or desktop study. It is also useful to refer to IStructE's *Assessing higher-risk buildings under the Building Safety Act: a compendium of structural typologies*[14.2].

Section 14.2.1 provides a list of potential sources of information and available tools for investigation.

14.2.1 Desktop study

The desktop study is the first investigation of the building, sometimes referred to as 'map regression'. These investigations can lead to a more complete picture of how the building was put together, what materials you may find on-site and how the site was previously used.

The historical context

A look back at the history of the site can give us clues to the site's original use and can be helpful in determining whether further detailed investigations are necessary. For example, if the building has potential archaeological merit in the ground, then any new excavations over the site footprint will need careful co-ordination with an archaeological specialist, and this may have some project risk associated with it.

Doing our homework here can save money and time in the project. Typical sources for the UK are:

- Listing (Historic England)[14.3]
- Checking if the building is in a conservation area or archaeological protection zone? Consult your local planning authority
- RIBA Archives/National Archives[14.4]
- Geotechnical maps — British Geological Society (BGS)[14.5]
- Flood maps[14.6]
- Bomb maps
- Mines[14.7]
- Railway
- Conservation architect (they may have produced a conservation plan/statement of significance)

Changes to the building over time

Some research can be caried out which aids our understanding of the changes which may have occurred to the building over its lifetime. Sources of interest are:

- UK planning portal or local building control — for as-built drawings or original building control applications
- National House Building Council (NHBC) (they only keep records for 15 years)
- Ordnance Survey maps — can tell us how a site has changed over the years. These maps are available online from subscription services (Figure 14.1)
- Conservation groups — local knowledge of the building. There may be historic groups which look after the building or parts of the building. They may hold records and be an invaluable source of information. Question these people as they will inevitably have lots of historic stories around the building and can fill in knowledge gaps. They are often very keen to share knowledge and save you a lot of time looking for information. Some historic groups hold documents or have produced technical information themselves. Sometimes they may be made up of retired engineers, if you're lucky!
- Old photographs/plans — historic photographs/plans are often extremely helpful to understand the previous use of the building. They also define when parts of the building may have been constructed and what parts were added at what time (Figures 14.2 and 14.3). This is an effective way to cross-check the representation of the buildings using the old maps which may have been sourced

Once some initial research has been carried out, the next step is to produce a set of existing drawings (these can be produced by hand but are more likely in CAD). It is good practice to commission a measured survey. Surveying companies can survey buildings very efficiently and this saves money for the client overall. The survey can act as a base for the existing drawings.

Figure 14.1: Example of historic Ordnance Survey map

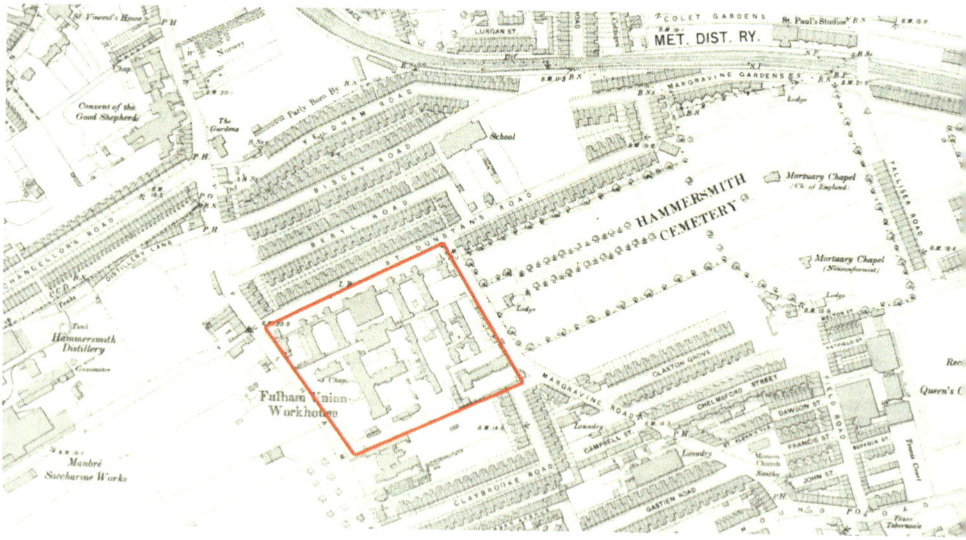

Figure 14.2: The Swan & Sugar Loaf (Croydon) was built in 1896 serving as a hotel. The site is current owned by Tesco, serving as an 'Express' store

Figure 14.3: Plan drawing, using colour to denote the different periods of construction

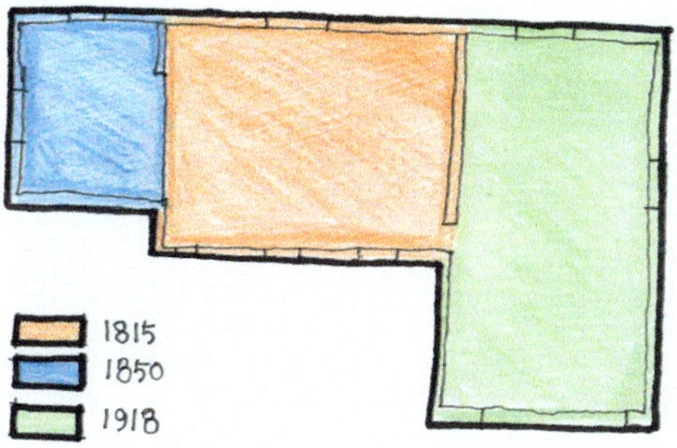

Once these drawings have been produced, they can be used on-site to record notes about the building. Consider drawing plans, sections through critical areas in the building and make a set of elevations. The plans are useful to record position and size of structure. The sections are helpful in noting how the typical structural form is set out, and elevations are useful to record material make-up and any structural movement in the facade. Consider if a 2D or 3D survey is of benefit to the work on the building. A 2D survey is the best place to start and then if there are some localised areas of concern a point cloud survey in 3D may have its uses for more detail. You may have located a set of existing/as-built plans of your building and these can be used on initial visits and form the basis of your existing set of drawings (Figure 14.4). Expect to find these drawings in different mediums based on when the building was designed and built. Some older work will be hand drawn, so only paper copies may be available. Some drawings will be on microfilm and more modern data will be stored electronically.

Microfilm was used historically to store a large number of photographed documents on a small slide. It required a special 'microfiche' reader to read the films[14.8].

Figure 14.4: Original building drawings found during a building control archive search

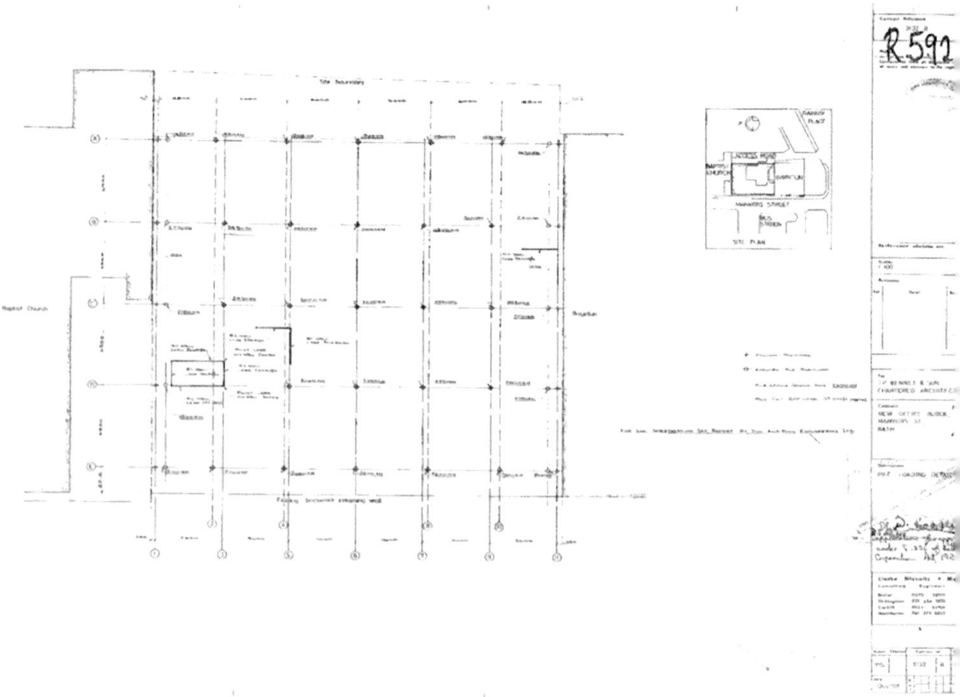

More advanced tools for surveying are available such as terrestrial scanning or drone surveys. With current developments in drone technology, an existing building can be visually surveyed with relative ease. This can be carried out quickly and safely and the results are instantly available. You can advise what you want recorded and this type of survey is cost-effective. It is, of course, limited to where the drone can fly and is non-invasive. It is particularly useful when access is limited — for example asbestos roofs or fragile roofs which limit a walk-over survey (Figure 14.5).

The work that a consulting engineer can do is to investigate the building extensively. However, further de-risking can be carried out at both the design and construction phases.

Figure 14.5: Aerial drone photo

Once a desk study has been carried out, further investigations may be required.

14.3 INSPECTION – HOW TO CARRY OUT A SITE VISIT

According to ISO 13822[14.1], an inspection is an 'on-site non-destructive examination to establish the present condition of the structure'. Once the desktop study is completed and, hopefully, a set of existing plans or drawing is available, an inspection on-site is the next milestone in the preliminary stages of a project of assessment and rehabilitation of an existing building.

You may be going for an initial look around with the client or another design team member. A crucial point to note — you should never visit the site alone! This is a personal risk to you, as potentially there are many hazards on-site such as trip hazards or weak structures. It is important to team up and visit the site together.

14.3.1 Preparing for a site visit

Every site will be different but there is always a basic list to bear in mind to support your visit. This is not exhaustive but is a good starting point:

- Carry out a risk assessment and have it approved by a senior member of your team before heading to site
- Check the asbestos report for all areas you are visiting (you will have collated this within the desktop study). Make sure opening up is carried out by a qualified asbestos team prior to your inspection visit. It is often unlikely that you will be able to access all areas and fully open up to expose all structure, so this may mean some of the opening up works are left until the commencement of the works by the contractor
- Check all equipment (including PPE) making sure you have high visibility jackets, hard hats, steel toe-capped boots. Even if you are visiting an existing building that is in use, being site-ready shows a level of professionalism (Figure 14.6)

Figure 14.6: Suggestions for site visit equipment

14.3.2 What to do when you get on-site

As mentioned previously, going to site with another person partially covers the safety aspect, however another pair of eyes is also useful and another brain is a good sounding board for what you find on-site. In the case of producing an in-depth site investigation of the building, one person can measure structural elements and the other can record on the drawings; this shortens time spent on-site and means the information is in two heads when you get back to the office.

Upon arrival have a brief overview of the site from outside and look for any major safety issues which can be seen from this vantage point, i.e., major movement in the structure. Ask yourself and share your thoughts with your colleague on whether it is safe to proceed with the internal survey. At this point you can record any facade issues on the existing elevations. When you enter the building always enter at ground floor (or basement level if there is one) and work upwards. From here you can check each floor above you from below before you stand on it!

Using a set of pre-made drawings, walk the site recording what you can see and access safely. If there are areas to record in detail (or areas that have been opened to explore), sketch and dimension elements of this existing structure. You may also consider taking samples if you require test data — this aspect may depend on whether specific *in situ* testing is planned to improve the knowledge level of the structure.

You should not rush this stage, as gleaning as much information will help with your understanding of how the building was originally put together and any alterations that have occurred over the years (this will be a chance to test out your theory on what you expected to find from your investigation using the desktop study). In older derelict buildings beware of any animals such as pigeons, rats or squirrels and their droppings!

With existing structures it is also important to understand likely failure modes. This allows the engineer to understand fully any areas where failure of a single element could lead to catastrophic collapse, and design any investigation works accordingly. Giving thought to consequence class and change of use is vital in redeveloping a building. Refer to IStructE's *Structural robustness and disproportionate collapse in buildings* for more details[14.9].

14.4 FURTHER INVESTIGATIONS TO CONSIDER

According to Annex B of the *New European technical rules for the assessment and retrofitting of existing structures*[14.10] which reflects the framework provided in Eurocode 8, Part 3[14.11], the knowledge of an existing structure can be divided into three main components:

- Geometry — geometrical properties of the structural system
- Details — detailing of reinforcement for reinforced concrete, connections between steel components, connections between floors and wind bracing systems, characteristics of mortar in masonry joints, etc.
- Materials — actual material properties (estimated either from the desktop study or from destructive and non-destructive *in situ* testing)

The desktop study and the preliminary inspection discussed in Sections 14.2 and 14.3 generally allow addressing of geometry and details. Material properties are mostly assessed with intrusive and non-intrusive investigations that can be considered as a further step to understand the building fully. These are material-specific and the more we understand all parts of the building in terms of material, load paths, structural layout, the less uncertainty we will take forward into the refurbishment/maintenance period.

A list of potential techniques available for simple on-site testing and material testing are provided in Chapter 5 of the IStrutctE's *Appraisal of existing structures*[14.12]. In particular, bespoke tables are provided for different construction materials: concrete structures, metal structures, masonry, timber and polymers and fibre polymer composites (Tables 5.1–5.6).

14.5 CONCLUSIONS

Currently, reuse of existing buildings is a niche area, but as more practices move their focus from new-build to reuse, the checking of existing buildings will become key. Engineering practices that are able to prove, to themselves and through calculation, that reduced, or no works are necessary for the change the client wants will become highly valued. Part of the success of this approach is through the gathering of data about the existing building and the use of it in a way that reduces uncertainty and increases the engineer's confidence in the design. In Chapter 15 we will look at another statistical approach that can be applied once we have data, especially quantitative data.

REFERENCES

14.1 *ISO 13822:2010: Bases for design of structures. Assessment of existing structures.* Geneva: ISO, 2010.

14.2 The Institution of Structural Engineers. *Assessing higher-risk buildings under the Building Safety Act: a compendium of structural typologies.* London: IStructE, 2024.

14.3 Historic England website for listed buildings. Available at: https://historicengland.org.uk/listing/the-list/ [Accessed: January 2025].

14.4 The National Archives website. Available at: https://www.nationalarchives.gov.uk/ [Accessed: January 2025].

14.5 British Geological Society (BGS) website. Available at: https://www.bgs.ac.uk/ [Accessed: January 2025].

14.6 HM Government. *Get Flood Risk Information for Planning in England.* https://flood-map-for-planning.service.gov.uk/ [Accessed: January 2025].
14.7 The Coal Authority. *The Coal Authority Map Viewer.* Available at: https://datamine-cauk.hub.arcgis.com/ [Accessed: January 2025].
14.8 Microfiche readers. Available at: https://en.wikipedia.org/wiki/Microform [Accessed: January 2025].
14.9 The Institution of Structural Engineers. *Structural robustness and disproportionate collapse in buildings (Second edition).* London: IStructE Ltd, 2023.
14.10 Luechinger P. *et al. New European Technical Rules for the Assessment and Retrofitting of Existing Structures.* January 2015. Available at: 10.2788/052881 [Accessed: January 2025].
14.11 *CEN (2005) EN 1998-3-1-4: Eurocode 8, Design of Structures for Earthquake Resistance, Part 3, Assessment and Retrofitting of Buildings.* Brussels: CEN, 1998.
14.12 The Institution of Structural Engineers. *Appraisal of existing structures (Third edition).* London: IStructE Ltd, 2010.

CHAPTER 15

Bayesian updating

Raffaele De Risi and Flavia De Luca

15.1 INTRODUCTION

In Chapters 13 and 14 we looked at the framework for assessing buildings and how we might go about collecting data about the building, both qualitative and quantitative. In this chapter we will take a slight detour, returning to the reliability theory of Chapter 6 and the probability distributions presented in Chapter 10 and looking at how we might apply Bayesian updating.

Bayesian updating allows combining and updating uncertainties as new data becomes available. While this approach has many applications (e.g., it can be employed to update preliminary designs of new buildings), in reuse, where some information is known but more information becomes available as we go through the design and build process, it is particularly helpful.

15.2 BAYESIAN UPDATING

When approaching an existing building the design is often based on information available from the desk study. This could include design loads from codes of practice available at the time of design, or from documented strengths provided on as-built drawings. The initial design is carried out with a specific characteristic or average value which represents prior information based on the standard values produced by the industry. In the case of strength, the assumed values should be checked during the next design phases, with experimental results on material samples collected on-site. However, only a small number of samples are available and this is indeed not sufficient to obtain a robust distribution for material properties with the possibility that the additional information does not reduce the degree of uncertainty in the design. However, this data can be used to update the prior information and come up with a posterior distribution of the strength that can be used to reassess the reliability of the structure after construction.

From a practical point of view, the Bayesian approach allows estimating the parameters of a distribution or the parameters of a regression. Specifically, the parameters of the distributions are treated themselves as uncertain, and the distributions of such parameters can be calculated using the following general approach[15.1]:

$$f(\theta|D) = c^{-1} L(D|\theta)f(\theta) \qquad 15.1$$

where θ represents the parameters of the distribution, D represents the new data, $f(\theta)$ is the prior distribution of the parameters θ, $L(D|\theta)$ is the likelihood function and represents the information contained in the observations and $f(\theta|D)$ is the posterior distribution describing the updated estimate of θ. The term c is a normalisation factor ensuring that the posterior's indefinite integral is equal to 1. The likelihood function depends on the adopted type of distribution and is proportional to the conditional probability of the data D given the parameter θ.

D here is denoted with an italicised *D* to differentiate it from the more generally used D used for load effect (or demand). The two should not be confused, but we have used them both rather than use another letter so that we conform with standard nomenclature.

Once the posterior probability distribution of the parameters $f(\theta|D)$ is computed, this can be used to define some predictive probability distribution of the generic random variable x for which we had the data D using the total probability law:

$$f(x) = \int f(x|\theta)f(\theta|D)d\theta \qquad 15.2$$

Note that the dependency on data D on the left-hand term of the equation has been omitted for the sake of notation simplicity.

Such a predictive distribution is available in closed form for some special cases[15.2].

15.2.1 Normal distribution with unknown mean and variance

Equation 15.1, for some cases, can be solved in closed form, as described in *Probability, statistics, and decisions for civil engineers*[15.3]. This is the case for normal distribution with an unknown mean and standard deviation. Let n' and n be the number of prior and new data, respectively. Also, let m' and m be the mean of prior and new data, respectively. Finally, let s' and s be the standard deviation of prior and new data, respectively. The mean of the posterior mean and standard deviation are:

$$\mu_{posterior} = m'' = \frac{n \cdot m + n' \cdot m'}{n''} \qquad 15.3$$

$$\sigma_{posterior} = s'' \cdot \sqrt{\frac{n''-1}{2}} \cdot \frac{\Gamma\left(\frac{n''-3}{2}\right)}{\Gamma\left(\frac{n''-2}{2}\right)} \qquad 15.4$$

where $\Gamma(\cdot)$ is the Gamma function[15.4] and n″ and s″ are provided by the following relationships:

$$n'' = n + n' \qquad 15.5$$

$$s'' = \sqrt{\frac{(n-1)\cdot s^2 + n\cdot m^2 + (n'-1)\cdot s'^2 + n'\cdot m'^2 - n''\cdot m''^2}{n''-1}} \quad 15.6$$

It is important to emphasise that Equations 15.3 and 15.4 are the mean values of parameters of the posterior distribution of the variable. In *Probability, statistics, and decisions for civil engineers*[15.3] you can find the whole posterior joint distribution of μ and σ. Moreover, these equations are valid if n″ is larger than 3.

Formulations for the predictive distribution of mechanical properties are available in Part 2 'Reliability updating and decision analysis' of the JCSS *Probabilistic assessment of existing structures*[15.5]. For this special case, the predictive distribution of the random variable is the following Student's t-distribution with $n'' - 1$ degrees of freedom of the following variable:

$$f(x) = T_{n''-1}\left(\frac{x - m''}{s''}\sqrt{\frac{n''}{n''+1}}\right) \quad 15.7$$

where $T_{dof}(\cdot)$ is the generic representation of the Student's t-distribution with a specific number of degrees of freedom (dof). In case of lognormal distribution, the distribution is a log-Student's t-distribution as discussed in *Updating the prior parameters of concrete compressive strength through Bayesian statistics for structural reliability assessment*[15.6].

15.2.2 Example

Let us consider again the case study presented in Chapter 6 and shown again for clarity in Figure 15.1. An engineer is concerned with distribution of the acting force F. Engineering judgement suggests that proper prior estimates of the mean and standard deviation coincide with the numbers proposed in Figure 15.2, i.e., m′ = 4kN and s′ = 0.5kN. These values were obtained by a previous experimental campaign on four similar structures, i.e., $n' = 4$.

Figure 15.1 (a) Demand and (b) capacity models

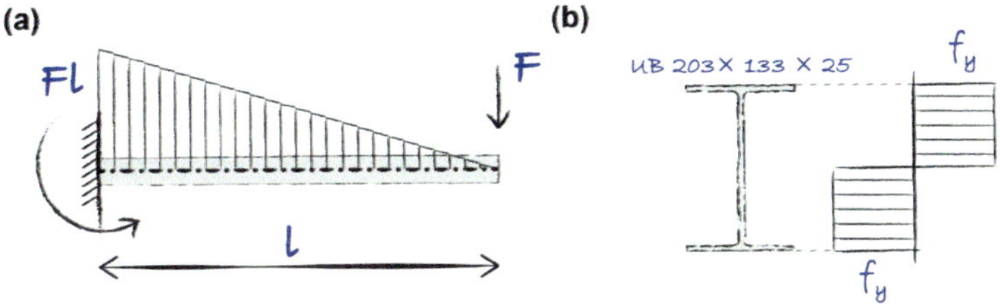

To update the uncertainty on the force, the engineer collects measurements on five similar structures, and the results are D = [2.1 2.7 2.8 3.1 2.75]kN. Following these five measurements (n = 5), the mean is m = 2.68kN, and the standard deviation is s = 0.365kN. The posterior mean and standard deviation are $\mu_{posterior}$ = 3.27kN and $\sigma_{posterior}$ = 0.96kN, with s″ = 0.80kN. Figure 15.2 shows the prior and posterior distribution of the force F based on the new data D. The posterior distribution of the variable is also the predictive model of the random variable (Equation 15.7) after the acquisition of the new data.

Figure 15.2: Prior and posterior distribution of F based on the new data

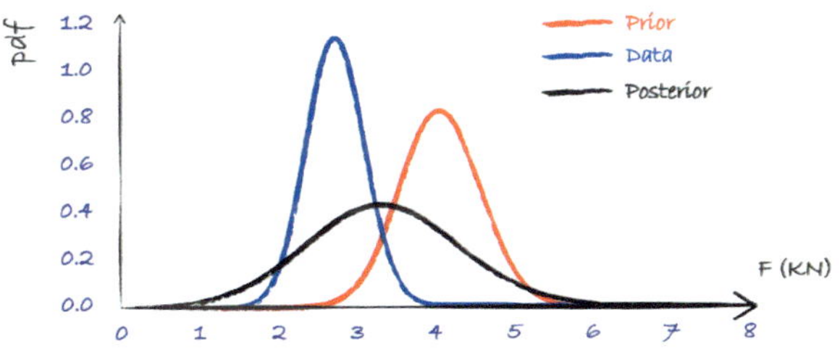

Using the posterior distribution for the reliability assessment in this example, it is possible to calculate the reliability index and the probability of failure after acquiring the new information. Specifically, $\beta = 2.90$ (previously $\beta = 2.77$) and $P_F = 1.8 \cdot 10^{-3}$. This number is still not acceptable according to the provided limits in Table 6.2 where for 50 years RC2 $\beta \geq 3.8$. In this case the new data is quite far from the prior distribution and the updating is leading to a posterior with a significantly higher standard deviation. Note that in this example the standard deviation was unknown and the additional data informed both parameters of the distribution.

15.2.3 Normal distribution with known variance and unknown mean

In some exceptional cases (e.g., steel rebars produced on a large scale), the variance σ'^2 is known, and only the mean of the posterior distribution is unknown. This is the case where the variance is provided, for example, by the producer of a specific structural component as a result of quality control. In this case, the parameters of the posterior distribution of the parameter of interest[15.7] can be obtained as:

$$\mu_{posterior} = \frac{\dfrac{m'}{\sigma'^2} + \dfrac{m \cdot n}{s^2}}{\dfrac{1}{\sigma'^2} + \dfrac{1}{s^2}} \qquad 15.8$$

$$\sigma_{posterior} = \sqrt{\sigma'^2 + \frac{s^2 \cdot \sigma'^2}{s^2 + n \cdot \sigma'^2}} \qquad 15.9$$

where m′ and m are the prior mean and the mean of the new data; *s* and *n* are the standard deviation and the number of the new data.

15.2.4 Update of probability distributions

To address the possible results of an updating process it is useful to show the different scenarios. This is proposed through the cases shown in Figure 15.3 adapted from *Probabilistic assessment of existing structures*[15.5] and *On reliability assessment of existing structures*[15.8]. Figure 15.3 shows cases of 'weak' and 'strong' distributions for the likelihood and the prior, respectively, and their effect on the resulting posterior distributions.

When the new data has high dispersion and is represented by a small sample size, the corresponding likelihood will be considered 'weak' and it will have a high standard deviation. Figure 15.3a shows a strong prior and a weak likelihood, Figure 15.3b a weak prior and a strong likelihood and, finally, Figure 15.3c shows a case of prior and likelihood distributions with comparable strengths.

Figure 15.3: Illustration of updating of probability distribution

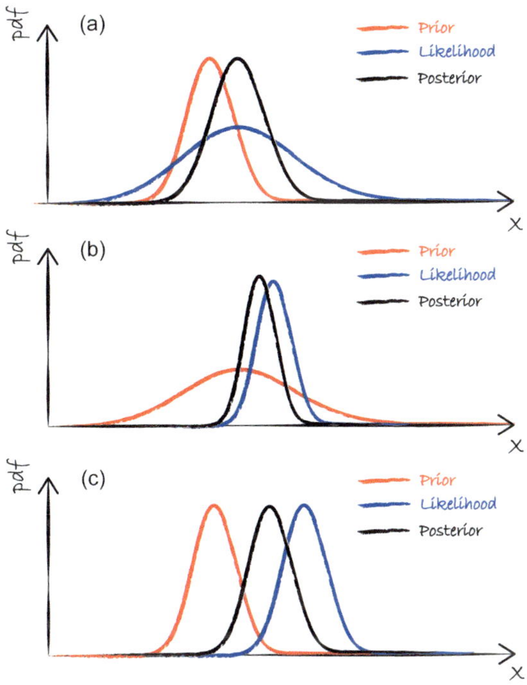

15.3 CONCLUSIONS

The structural engineering industry, and more broadly the construction industry, being risk-averse by nature, has been slow in integrating data techniques in conventional design approaches. For example, it would be great to have sensors measuring the occupancy levels (both static and dynamic) in buildings so that a reliability profile could be computed in real-time along the life cycle of the structure and new data could be gathered for the design of future buildings. The

Bayesian approach is the perfect framework to incorporate new data in the initial estimations. Such an approach will also help to combine rationally the deterministic values proposed by the codes, with objectively measured data to improve the design assumptions systematically. It is clear that improving standards, and challenging historical assumptions, is essential if we are to improve carbon efficiency in line with the net zero agenda.

In Chapter 16 we will look at the quantitative assessment of existing buildings.

REFERENCES

15.1 Box, G.E.P. and Tiao, G.C. *Bayesian Inference in Statistical Analysis*. New York: Wiley, 1973.

15.2 Rackwitz, R. 'Predictive distribution of strength under control'. *Matériaux et Constructions*, 16, 1983, pp259–267. Available at: https://doi.org/10.1007/BF02473691 [Accessed: January 2025].

15.3 Benjamin, J. R. and Cornell, C.A. *Probability, Statistics, and Decisions for Civil Engineers*. New York: McGraw-Hill, 1970.

15.4 Artin, E. (2015). *The Gamma Function — Dover Books on Mathematics*. New York: Dover Publications Inc., 2015.

15.5 Joint Committee on Structural Safety. *Probabilistic Assessment of Existing Structures*. RILEM, 2001. Available at: https://www.rilem.net/publication/publication/96 [Accessed: January 2025].

15.6 Feiri, T. *et al.* 'Updating the Prior Parameters of Concrete Compressive Strength Through Bayesian Statistics for Structural Reliability Assessment'. *Structures*, 58, December 2023. Available at: https://doi.org/10.1016/j.istruc.2023.105636 [Accessed: January 2025].

15.7 Faber, M.H. 'Statistics and Probability Theory'. *Topics in Safety, Risk, Reliability and Quality*, 18, 2012.

15.8 Diamantidis, D. *et al.* 'On Reliability Assessment of Existing Structures'. *Structural Safety*, February 2024. Available at: https://doi.org/10.1016/j.strusafe.2024.102452 [Accessed: January 2025].

CHAPTER 16

Quantitative assessment

Flavia De Luca

16.1 INTRODUCTION

In this final chapter on reuse, before we present a case study, we discuss the different methodologies available for quantitative assessment, referring to current available guidance. It should be noted that this is a rapidly-expanding area and we should expect frequent new and updated guidance as confidence and experience in this space grows.

16.2 ASSESSMENT BY CALCULATION OF EXISTING STRUCTURES

In all the situations in which the preliminary assessment does not lead to the conclusion that the structure is reliable for its remaining working life, a detailed assessment can be required (ISO 13822[16.1]). According to CEN/TS 17440[16.2] the assessment of an existing structure can be performed 'by calculation' or 'based on past performance' subject to conditions being met. The use of testing and monitoring can support and be part of either approach.

Assessment by calculation requires:

- Definition of basic variables
- Structural modelling and analysis
- Selection of the verification method

The aspect of definition of basic variables, according to CEN/TS 17440, covers knowledge of the structure and the different tools discussed in Chapter 14, to acquire sufficient information with specific indications on how to deal with typical loads and actions (e.g., self-weight, imposed loads, snow, wind, etc.) and the assumptions for materials.

In situ tests on materials can be used as part of an updating procedure with respect to prior information produced from other sources. An update procedure is explicitly discussed in both the CEN/TS 17440 and ISO 13822. Updating was covered in Chapter 15 of this book; the possibility of using it in the assessment of existing buildings is crucially important and it is explicitly discussed in annexes of codes and recommendations — with some examples proposed in the JCSS document on *Probabilistic assessment of existing structures*[16.3] and studies assessing the impact on recent code developments[16.4].

The development of a structural model for an existing structure is, again, characterised by very specific assumptions made on static and kinematic boundary conditions, and by other fundamental aspects of the model that can significantly affect the output. Setting up a building information model (BIM) early in the design process can allow coherent collection of all information in a single environment, to develop the specific structural model required to assess the capability of the structure to carry loads.

16.3 KNOWLEDGE LEVEL DEFINITION FOR EXISTING STRUCTURES

The selection of structural models and analysis methodology can be refined as soon as more detailed information is available. This is also reflected in the concept of achieving a specific 'knowledge level' for the structure that makes preparing a more sophisticated model (e.g., a non-linear model)[16.5] more meaningful.

For example, in Table 3.1 of Eurocode 8, Part 3 , a 'normal' knowledge level is the minimum requirement to develop a non-linear model for the structure. With increased knowledge of the structure, the use of a lower confidence factor is allowed in Eurocode 8, Part 3. Such confidence factors are used in addition to typical material safety factors, explicitly recognising the additional level of uncertainty with an existing structure. These aspects lead directly to the overall consideration that an additional investment into inspections and testing (i.e., knowledge level) allows a more detailed assessment and higher confidence in the design of any retrofitting measure.

It is common in professional practice that a preliminary model is built based on very basic information obtained from a first survey or even from the desk study, and then the model is improved and detailed based on further information, evidence or testing. In this context, it is hugely important to understand the original methods of analysis used. This can be captured through the study of manuals and code recommendations that were current at the time of the original design.

The major barrier for extensive testing is often the necessity to avoid disruption in the use of the building and to prevent damage. This can sometimes lead to samples not being taken in the specific area of interest for the engineer, or other necessary compromises e.g. the level of invasiveness that can be achieved through testing. Another option is to use non-destructive testing (NDT) techniques. NDT techniques typically use previously calibrated relationships for the assessment of material properties in existing buildings[16.6]. A valuable suggestion is to have an *ad hoc* adjustment of NDT relationships using data from the existing structure for which some destructive tests are available. NDT measurements can be taken at the location of the destructive tests to update the empirical relationship for material strength prediction, accounting for the specific conditions of the structure at hand[16.7–16.9]. Refer also to Chapter 17 in this book.

16.4 VERIFICATION METHODS FOR EXISTING STRUCTURES

There are four verification methods for existing structures that broadly increase in complexity:

- Partial factor method (as for newly-designed structures)
- Assessment value method
- Probabilistic method
- Risk assessment method

The suitability of each (and choice of method ultimately selected) can be discussed with the client. The conceptual bases of the first three have been covered in Chapter 6, and the final method is the focus of Part 5 of the book.

The additional aspect to be considered with existing buildings is that the acceptable level of risk can be different from a new construction, and it can involve the ALARP principle (covered in Part 5). The regulatory context for existing buildings in the UK is reviewed in Appendix 2 of IStructE's *Appraisal of existing structures*[16.10]. This Appendix covers the legal position in the UK for acceptable risk levels. Legal criteria vary significantly depending on the situation and the planned use of the building after any intervention. ALARP is generally applied to existing building assessment. Where replacement work is carried out, the structure must comply with building regulations and the 'non-worsening' principle applies to unaffected parts of the building. In case of change of use for the entire building, the structure must comply with building regulations.

CEN/TS 17440 provides the indications of the specific equations to implement each of the approaches, with a rich literature of examples available in various publications[16.3,16.11,16.12]. Note that the first attempt should always be to use the partial factor method, assessing the existing structure as if it was a new one. Then, the other methodologies are implemented as solutions to overcome different barriers, and among those 'the conservatism of partial factor methods'.

As a tool to be used only in very specific situations, the 'assessment based on past performance' is allowed by the CEN/TS 17440 even if its use is devolved to National Annexes. In all cases in which this approach is employed, more emphasis is put on inspection and suitable level of knowledge of the structure. This is mainly because the inspection itself is used to assess the absence of evidence of significant damage, distress, deterioration. From a reliability point of view, the satisfactory performance of a structure during a specific known period of years (T) indicates that resistance is higher than load for this period given the absence of deterioration and damage (Annex B of CEN/TS 17440).

While from a methodological point of view, the approaches for verification in existing structures also represent an option in the case of new design, the real aspect where different judgement is necessary is related to reliability targets. In particular, the different reference period is the first factor influencing the assumption of different target values (Table 13.3) and, in general, the target for an existing building can be lower than that of a new structure, as the relative cost of safety measures to increase reliability of an existing building is greater than that of a new structure (Annex C of CEN/TS 17440).

16.5 CONCLUSIONS

We have now covered reuse from various perspectives, from the framework to facilitate checking existing buildings, to the investigation of these buildings, communication with clients, quantitative assessment and the role of Bayesian updating. Finally, we will present a case study which shows the future potential of these approaches.

REFERENCES

16.1 *ISO 13822:2010: Bases for design of structures. Assessment of existing structures.* Geneva: ISO, 2010.

16.2 *CEN/TS 17440:2020: Assessment and Retrofitting of Existing Structures.* Brussels: CEN, 2020.

16.3 Joint Committee on Structural Safety. *Probabilistic Assessment of Existing Structures.* RILEM, 2001. Available at: https://www.rilem.net/publication/publication/96 [Accessed: January 2025].

16.4 Diamantidis, D. *et al.* 'On Reliability Assessment of Existing Structures'. *Structural Safety*, February 2024. Available at: https://doi.org/10.1016/j.strusafe.2024.102452 [Accessed: January 2025].

16.5 *CEN (2005) EN 1998-3-1-4: Eurocode 8, Design of Structures for Earthquake Resistance, Part 3, Assessment and Retrofitting of Buildings.* Brussels: CEN, 1998.

16.6 Breysse, D. (2012). 'Nondestructive Evaluation of Concrete Strength: An Historical Review and a New Perspective by Combining NDT Methods'. *Construction and Building Materials*, 33, August 2012, pp139–163. Available at: https://doi.org/10.1016/j.conbuildmat.2011.12.103 [Accessed: January 2025].

16.7 Huang, Q. *et al.* 'Predicting Concrete Compressive Strength Using Ultrasonic Pulse Velocity and Rebound Number'. *ACI Materials Journal*, 108(4), July 2011, pp403–412. Available at: https://doi.org/10.14359/51683113 [Accessed: January 2025].

16.8 Rainieri, C. *et al.* 'Non-destructive Characterization and Dynamic Identification of a Modern Heritage Building for Serviceability Seismic Analyses'. *NDT & E International*, 60, December 2013, pp17–31. Available at: https://doi.org/10.1016/j.ndteint.2013.06.003 [Accessed: January 2025].
16.9 Pucinotti, R. 'Reinforced Concrete Structure: Non Destructive *in situ* Strength Assessment of Concrete'. *Construction and Building Materials*, 75, January 2015, pp331–341. Available at: https://doi.org/10.1016/j.conbuildmat.2014.11.023 [Accessed: January 2025].
16.10 The Institution of Structural Engineers. *Appraisal of existing structures (Third edition)*. London: IStructE Ltd, 2010.
16.11 Ellingwood, B.R. *Structural Safety and Its Quality Assurance — Chapter 9*. Reston: American Society of Civil Engineers, 2005.
16.12 Holický, M. *et al.* 'Reliability Basis for Assessment of Existing Building Structures with Reference to SANS 10160'. *Journal of the South African Institution of Civil Engineering*, 63(1), March 2021. Available at: https:/doi.org/10.17159/2309-8775/2021/v63n1a1 [Accessed: January 2025].

CHAPTER 17

Case study

Flavia De Luca

17.1 INTRODUCTION

In this case study we will look briefly at how, through analysis and testing of materials sourced from the building, the Tower of the Nations was checked for earthquake loading, which only became a consideration in the area after the building's construction.

17.2 TOWER OF THE NATIONS

The Tower of the Nations is a modern heritage structure located in Naples. The building was designed in 1938 by Venturino Venturia and Carlo Cestelli Guidi as architect and engineer responsible for the project. The building was completed in 1940 (Figure 17.1). After WWII the

Figure 17.1. The Tower of the Nations (under construction, and in completed form)

building fell into disrepair, and it was then included in a variety of requalification plans for the entire park in which it is located — the Mostra d'Oltremare complex in Naples.

This is a typical case of a structure designed for gravity loads only, and then needing to be reassessed for earthquake loads, since the area of Naples was later characterised as a seismic zone in 1981 following the destructive Irpinia earthquake which affected the city.

The structure has since been the object of a detailed assessment study developed in accordance with Eurocode 8, Part 3[17.1] and for which a non-linear analysis method was performed[17.2,17.3].

The framework for definition of knowledge level provided in Eurocode 8, Part 3 and also discussed in *New European technical rules for the assessment and retrofitting of existing structures*[17.4] is applied to this structure. Notwithstanding its modern heritage nature, the building was inspected with invasive techniques in specific parts to assess the quantity and dimension of structural reinforcements. The status of the foundations was investigated through a specific inspection at basement level (Figure 17.2). The remaining part of the reinforcement details was developed through a simulated design procedure relying on the structural design code employed at the time[17.5].

For material testing a combination of destructive and non-destructive methods were employed (Figure 17.3). The assessment of concrete compressive strength was obtained through a number of destructive compressive samples and then using such samples for *ad hoc* calibration of non-destructive method (i.e., sonic rebound method).

The importance of the structure, its nature as a modern heritage structure and the funding available for the

Figure 17.2: Example of *in situ* inspections performed on the Tower of the Nations

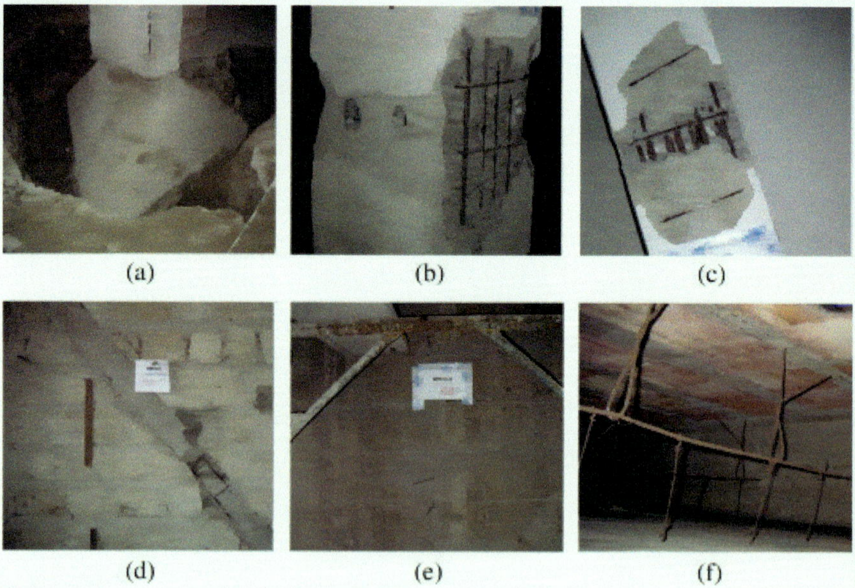

Figure 17.3: *In situ* testing performed on the Tower of the Nations

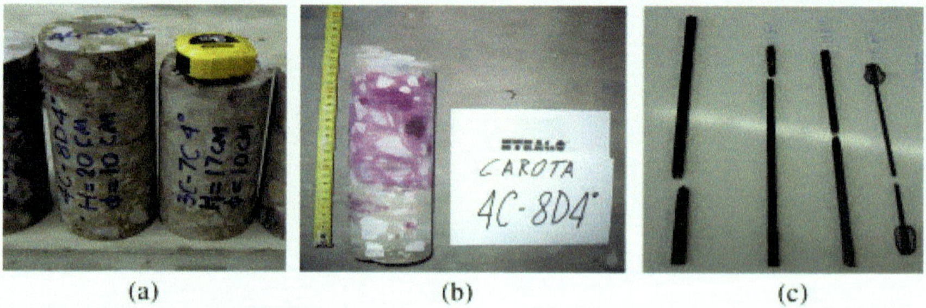

investigation allowed a knowledge level of two to be achieved ('normal' knowledge level) as defined in Eurocode 8, Part 3[17.1]. This knowledge level meant the use of a confidence factor of 1.2 for the assessment. This confidence factor was employed to reduce the value of average compressive strength for concrete and average steel yielding strength obtained from the *in situ* testing. The limiting factor to achieve the maximum level of confidence was the number of destructive tests available on the concrete. The confidence factor, as coefficient, is used to reduce the average compressive strength and considers the fact that in existing structures *in situ* tests provide average compressive strength not characteristic values.

Eurocode 8, Part 3 requires three samples per floor to achieve full knowledge of the structure corresponding to the use of a confidence factor equal to 1.0 (Table 17.1). This represents a prescriptive requirement that, while it provides suitable guidance to design and *in situ* inspection and testing, it can also represent a barrier in cases where it is not possible/feasible to take samples at all floors, limiting the knowledge level and imposing significantly conservative assumptions through the confidence factor allowed for use. Different considerations can be followed for other structures when the prescriptive aspect of Eurocode 8, Part 3 are not relevant as, for example, in the UK. Still, some quantitative consideration about a minimum number of samples should be made.

Table 17.1: Recommended minimum requirements for different levels of inspection and testing for reinforced concrete structures

	Inspection (of details)	Testing (of materials)
	For each type of primary element (beam, column, wall):	
Level of inspection and testing	Percentage of elements that are checked for details	Material samples per floor
Limited	20	1
Extended	50	2
Comprehensive	80	3

Concrete cylindrical average compressive strength (f_{cm}) for this structure was evaluated as equal to 16MPa from a combination of non-destructive tests and from concrete specimens (Figure 17.3). This is not too different from the code prescriptions for material properties according to which the structure was originally designed[17.6,17.7], and it shows how former building codes employed at the time of the original design can become a suitable source of prior information, and *in situ* tests can then be used to update this information (Chapter 15).

17.3 ASSESSMENT OUTCOME

The seismic assessment of the structure proved the existing building performance was satisfactory, when assessment was made using an advanced non-linear dynamic analysis for both damage limitation and life safety limit state, using seven earthquake acceleration time-histories as seismic input. This outcome was achieved by *in situ* inspections and testing which allowed the designers to reach a knowledge level two for the structure. This was combined with an advanced non-linear analysis method to assess the performance and account for non-linear behaviour of materials. The structural performance of the building was assessed based on interstorey drift and element chord rotation assumed as engineering demand parameters. If the level of knowledge achieved had not been at least level two, the use of such an advanced analysis method would not have been allowed by the prescriptions of Eurocode 8, Part 3.

The performance also proved to be satisfactory when the structural contribution of non-structural elements, such as tuff masonry infills, was considered. Tuff is a relatively soft volcanic stone[17.8]. The result is not surprising given the likelihood that the design already considered lateral stability of the structure under wind load at the time of construction.

Figure 17.4 presents the overall performance of the Tower of the Nations showing maximum interstorey drift (M IDR) when subjected to damage limitation (DL) and significant damage (SD) earthquake demand as per Eurocode 8, Part 3. Earthquake intensity is expressed in terms of spectral acceleration ($S_a(T_{el})$)[17.2].

Figure 17.4: Overall performance of the Tower of the Nations in terms of intensity measure and engineering demand parameter for the bare model in (a) X and (b) Y direction and infilled model in (c) X and (d) Y direction, where X corresponds to the glazed facade and Y corresponds to the infilled facade in Fig. 17.1 (right)

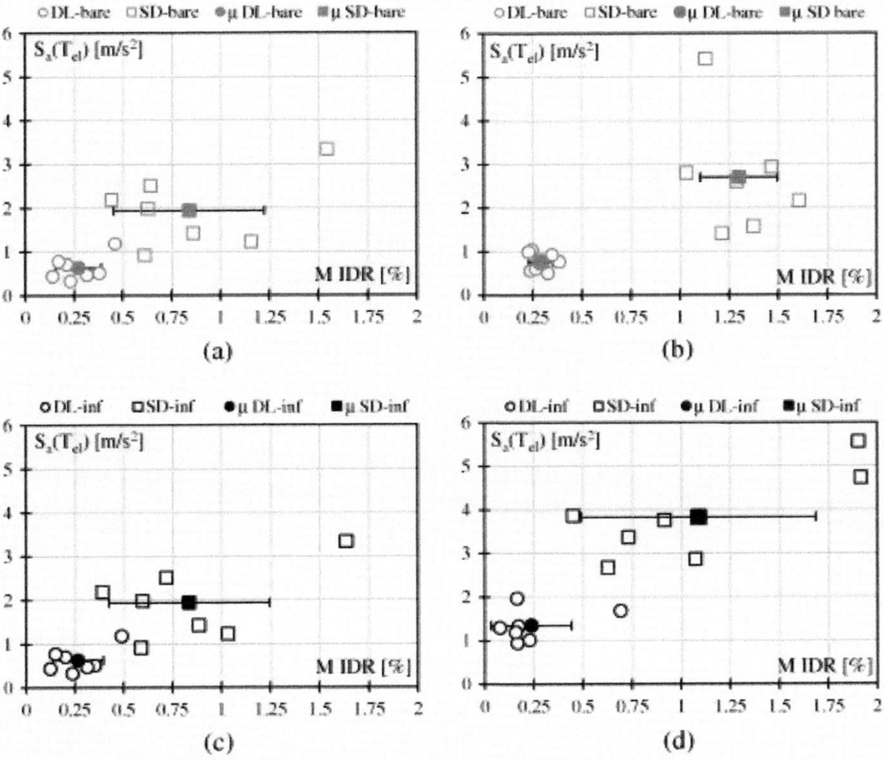

The median value of maximum interstorey drift, assessed considering seven earthquakes for significant damage (i.e., 475 years return period) was slightly higher than 1.25% for the bare structure and well below this threshold (~1.1%) when tuff infills were considered (Figure 17.4). For damage limitation (i.e., 225 years return period), the demand in terms of IDR was within the capacity and not too different when the structural contribution of infills was considered. The consideration of infills indeed changed the damage distribution among storeys. The assessment results obtained using the non-linear dynamic analysis approach suggested by Eurocode 8, was also employed to obtain the fragility curve of the structure[17.3].

17.4 CONCLUSIONS

While this case study represents a specific situation; the checking of a building for earthquake loading following reclassification of the district in which it was built, it can be seen that carefully designed and delivered assessment of existing buildings is a powerful tool. It also demonstrates the important role an engineer can play in minimising structural intervention through careful calculation and the gathering of the right data using physical testing.

REFERENCES

17.1 *CEN (2005) EN 1998-3-1-4: Eurocode 8, Design of Structures for Earthquake Resistance, Part 3, Assessment and Retrofitting of Buildings.* Brussels: CEN, 1998.

17.2 De Luca, F. *et al.* Eurocode-based Seismic Assessment of Modern Heritage RC Structures: The Case of the Tower of the Nations in Naples (Italy). *Engineering Structures*, 74, September 2014, pp96–110. Available at: https://doi.org/10.1016/j.engstruct.2014.05.015 [Accessed: January 2025].

17.3 Verderame, G.M. *et al.* 'Seismic Assessment via EC8 of Modern Heritage Structures: Knowledge of the Structure and Analysis Methodologies'. *Handbook of Research on Seismic Assessment and Rehabilitation of Historic Structures*, June 2015, pp607–628), IGI Global. Available at: https://doi.org/10.4018/978-1-4666-9619-8.ch043 [Accessed: January 2025].
17.4 Luechinger P. *et al. New European Technical Rules for the Assessment and Retrofitting of Existing Structures.* January 2015. Available at: https://doi.org/10.2788/052881 [Accessed: January 2025].
17.5 Verderame, G.M. *et al.* 'A Simulated Design Procedure for the Assessment of Seismic Capacity of Existing Reinforced Concrete Buildings'. *Advances in Engineering Software*, 41(2), February 2010, pp323–335. Available at: https://doi.org/10.1016/j.advengsoft.2009.06.011 [Accessed: January 2025].
17.6 Decreto-Legge, R. 'Prescrizioni Regolamentari Italiane per l'Esecuzione delle Opere in Conglomerato Cementizio'. *Gazzetta Ufficiale*, 224, July 1933 [in Italian].
17.7 Regio Decreto 16/11/1939 n. 2229. 'Norme Per l'Esecuzione delle Opere in Conglomerato Cementizio Semplice e Armato'. *Gazzetta Ufficiale,* 92, April 1940 [in Italian].
17.8 Tufa and Tuff. *Designing Building* website. Available at: https://www.designingbuildings.co.uk/wiki/Tufa_and_tuff [Accessed: January 2025].

Part 4 Conclusion — Reuse

The reuse of buildings and building components is becoming an important part of any structural engineer's role, as we transition from new-build towards circular approaches to design. While this part does not attempt to act as a 'how to' it does highlight the potential importance of a deeper understanding of probabilistic design and reliability methods which go beyond the current code and design guidance.

As we reach the conclusion of Parts 2–4 we have highlighted the future potential of a deeper understanding of reliability-based design. Whether in designing to the code ...and no more or a regenerative design, where the classification and testing of materials requires understanding of different distributions and testing methods, or in reuse, where reliability analysis, coupled with Bayesian updating and testing can enable engineers to make better-informed decisions.

However, we have also highlighted the important role that the engineer must play. Leading the design, making important decisions, guiding the process and most importantly bringing their experienced judgement to ensure the design is safe. By combining these important elements, the future of engineering design feels both exciting and substantially different from the past.

But where does this all end? What is the future of design, as computing power increases, and we become more confident as designers using probabilistic tools? That is the subject of the final part of this book.

Part 5
The future of structural design

Structural design has been in a reasonably steady state for decades — but we know that change is coming. From increasing computing power that will be able to perform calculations that engineers have typically carried out, to having to manage a diverse set of risks at both local and global levels, the challenge is going to become more complex, and the tools we use will require:

- A better understanding of what design is trying to achieve
- What rules can be applied and when
- When the codes and regulations we have been applying are no longer applicable
- What the most suitable form of analysis is
- How we manage an ever-growing set of uncertainties as the risks we design for become more complex

Throughout the first four parts of this book, we have introduced different forms of reliability analysis by framing them around designing to the code and no more, and regenerative design practices, which lead to new and local materials and reuse.

The uncertainties that we have been looking at fall within three different levels of reliability-based design:

Level I: uncertainties are modelled by a characteristic/ nominal value, and structural safety is guaranteed via partial safety factors provided by building codes.

Level II: the uncertain parameters are modelled by the statistics of the random variables (e.g., mean and standard deviation).

Level III: the entire joint distributions of the uncertainties are considered.

However, a fourth level exists:

Level IV: all the uncertainties are considered in a fully-probabilistic approach, accounting for consequences of failure (e.g., losses, casualties).

As stated in Chapter 2, this fourth level is known as 'risk-based design'. Rather than simply aiming for an acceptable probability of failure, it now integrates the consequence of failure into the analysis.

Risk-based approaches are not novel, but they are currently the domain of specialists and are utilised in disciplines such as earthquake and offshore engineering.

In the future of design, as the risks become more complex and computing power increases, a risk-based approach, integrated with a deep understanding of engineering principles and experience, could enable engineers to unlock safer solutions for people and the planet.

But what would this look like, and how do we approach risk-based design? In this final part, we explore these questions.

CHAPTER 18

Risk-based design and risk acceptability

Raffaele De Risi

18.1 INTRODUCTION

As we enter the final part of this book, we will unpack what a quantitative approach to risk-based design looks like by building up all the constituent parts. For practising engineers, this approach is already applied in the form of qualitative risk assessments (or semi-quantitative approaches, where scores are given based on qualitative descriptions). It is important to clarify that this approach can be applied if all risk components are understood and quantifiable.

18.2 INTRODUCTION TO RISK-BASED DESIGN

Unlike other sectors, structural engineering is not a mass-production product-oriented field; it is about designing one-of-a-kind structures or infrastructure. This means that the load effect (structural demand) and capacity are highly variable, and uncertainties should be correctly included in the final design. Therefore, the final design should be associated with the risk, i.e., the frequency/probability of a failure and its consequence. A structural failure may have implications/consequences at multiple scales beyond the structure. Therefore, it is essential to consider the concept of risk assessment of a system involving the structure. Such a system requires detailed identification before design[18.1].

According to *ISO 2394 — General principles on reliability for structures*[18.2], structures should be designed considering their entire life cycle, support societal functionality, and enhance sustainable societal development. Sustainability is associated explicitly with performance levels such as life safety, environmental quality, cost efficiency, CO_2 emissions, loss of biodiversity, consumption of natural resources and energy utilisation. The sustainability aspect is key considering the current climate and biodiversity emergency[18.3]. *ISO 2394* also states that structural design should account for risk and is realised with:

- Building structures that function adequately under all expected actions (stressor), providing service and functionality during their life cycle
- Building structures that can withstand permanent, frequently repeated and extreme actions without damage or failure
- Building robust structures that do not suffer severe/cascading damage due to unforeseen/extraordinary conditions

A risk-based design consists of designing structures associated with a socially-acceptable risk level. This leads to consideration of whether they are acceptable and/or tolerable (Chapter 3) — which is not something we can simply answer in this book but should be an on-going conversation with policymakers as we develop this complex space. These non-trivial tasks can be conducted, for example, by comparing the calculated risk with other societal risks by looking mainly at empirical data and previous experience. Such an approach is not necessarily sufficient, however, past data and experience are the only things we can rely on at present. For example, the risk in terms of loss of lives per year is 10^{-3} for smoking, 10^{-4} for car accidents or construction works, 10^{-5} for air/train travel or building fire, and finally, is lower than 10^{-6} for structural failure[18.4]. It is evident that it is necessary to define maximum 'acceptable risk levels'. The definition of acceptable risk levels has a significant political dimension. Without political direction, technical representatives of society such as public authorities and building code developers, must come up with acceptable levels based on the combination of consequences and frequency/probability of events[18.5]. The societal consensus should be at several levels:

- Nature of the failure
- Acceptable individual risk associated with the individual person
- Acceptable collective risk associated with the entire society (dependency and interdependency of the impact chains)
- Risk transferred to future generations
- Distinction between risk to persons and risk to properties
- Distinction between natural and human-made risk
- Distinction between voluntary or involuntary risk
- Distinction between risk to a specified or unspecified group of people
- Distinction between risk characteristics that are well-known or unknown
- Acceptable costs associated with safety measures

Individual risk is the probability of the individual being harmed by a hazardous situation, and it depends on how voluntarily an individual undertakes a risky activity, and on the capacity of the individual to influence the risk by appropriate behaviour. Therefore, individual risk is key in risk perception. However, individual risk is not a primary concern from a societal perspective (i.e., decision-makers' perspective). Collective (i.e., societal) risk is of more interest. Society is more interested in the frequency/probability and severity of failures due to structural damage, leading to consequences for groups of people in economic, social and environmental terms. Therefore, a correct definition of risk limits should be based on all components mentioned previously, quantifying losses in monetary terms (e.g., reconstruction costs, interruption of economic activities), loss of lives or injuries, and impact on the environment (e.g., emissions to the atmosphere) both in the short- and long-term. Also, the impact on infrastructure should be considered as affecting recovery. The risk limit will represent a constraint for the optimisation necessary for a risk-based design (Figure 18.1). It is important to acknowledge that there may be cases where the optimal design is not acceptable (Chapter 7).

It is straightforward to compare losses when they are expressed with similar units, e.g., economic terms, environmental terms such as CO_2 emissions and loss of lives. However, it is difficult to define the acceptability of losses that are not presented in economic terms. The principle of the marginal lifesaving costs (MLSC) specifies that decisions leading to life safety are acceptable if the costs associated with possible means of saving one additional

Figure 18.1: Maximisation of the benefit due to the design optimisation

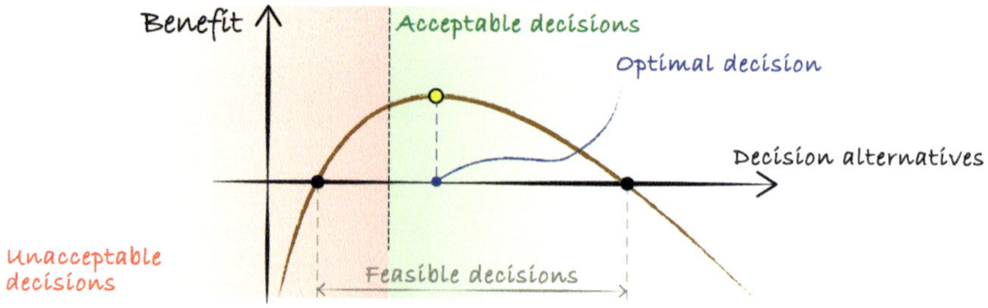

life are in balance with the societal willingness to pay for saving one statistical individual[18.2]. MLSC is definable within the ALARP framework (As Low As Reasonably Possible/Practicable), or with the SFAIRP framework (So Far As Is Reasonably Practicable)[18.6], and it can be adequately implemented using the life quality index (LQI)[18.7]. Regarding environmental requirements, absolute values for the qualities of the environment, usage and commitment of natural resources and emissions to the atmosphere are not easy to define[18.8]. However, these can be studied comparatively among the alternatives. On the environmental side, a reference parameter could be the damaged area in terms of terrestrial, freshwater and marine habitats, groundwater reservoirs, and other natural primary commodities[18.9]. An interesting approach is proposed by the NORSOK standard[18.10]; the environmental damage is categorised according to the average recovery time varying from six months for minor damage to 20 years for serious damage. Further suggestions on how to account for climate change in risk assessment are provided in the UNDRR technical guidance[18.11].

Finally, it is essential to clarify that full-risk design procedures are different from reliability-based design and can be replaced by reliability-based design procedures only if all the consequences of failure and damage are well understood and within normal ranges[18.2]. Further exemplification using semi-probabilistic modes can be used if, in addition to the consequences, failure modes and uncertainties can be categorised and standardised[18.2]. Therefore, a risk-based design should consider/anticipate all possible decisions and all possible future consequences of any potential failure/damage, using a combination of generalised life cycle cost (LCC) assessment and life cycle analysis (LCA) framework within an optimisation exercise similar to a traditional cost-benefit (CB) analysis or return on investment (ROI) analysis. Therefore, it is clear that risk-based design is necessary to deliver proper holistic multi-criteria risk-informed decision-making. A holistic multi-criteria risk-informed decision-making process is at the heart of donut economics, systems thinking and regenerative design.

18.3 RISK ANALYSIS

For a rigorous risk-based design, a risk analysis is needed. A risk analysis is not a straightforward process but rather an iterative exercise (Figure 18.2) that aims at estimating, appraising and monitoring the risk associated with a new structural design, and for changes in the use of existing structures and for significant modifications to an existing structure. The cyclic nature of risk analysis is on top of the conventional structural design that is (i) cyclic (Figure 18.2), as the same structure should be reassessed if changes are necessary to make the structural system work, (ii) repetitive, as all structural elements should be assessed with the same level of precision and rigour, and (iii) evolutionary, as the target of the engineer is to find the optimum[18.12]. All steps of proper risk analysis should always be coupled with adequate communication and thorough documentation

Figure 18.2: General framework for design-oriented risk analysis

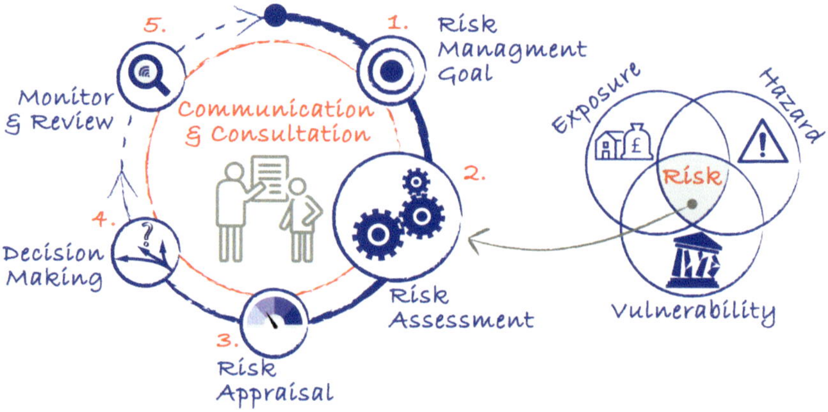

that is traceable and transparent, and through consultation with stakeholders. The reader will notice the similarity with many qualitative risk assessment processes.

18.3.1 Risk management goals

The first step of the risk assessment consists of defining the risk management goals that should reflect decision-makers' and societal values. A general discussion on risk criteria aligned with risk management goals was provided in Section 18.2. It has been emphasised that risk limits are defined comparatively with other societal risks derived through empirical data (i.e., previous failure events) and has a strong political dimension. Therefore, risk goals are targets against which risk will be appraised.

Examples of risk management goals are identifying the maximum tolerable expected probability of failure, the maximum tolerable expected economic annual loss, the maximum tolerable expected annual casualties/injuries, and the maximum tolerable average tonnage of CO_2 emission. This list is not exhaustive but provides an idea of possible risk goals and associated risk criteria to account for in the design.

These goals, and the associated criteria, are used later in the risk appraisal stage to determine if the risk is acceptable and, therefore, if risk mitigation strategies could be implemented. Such criteria are not static entities but rather dynamic factors that can be adjusted and tailored to the problem at hand following the outcomes of the first risk analysis cycle and following the subsequent monitoring/review operations and constant communication and consultation with stakeholders.

18.3.2 Risk assessment

This step consists of the systematic approach for describing and calculating risk. Risk is a measure of the combination (usually the convolution) of the hazard, the exposure and the vulnerability. The hazard is a stressor source (action) or a situation that may lead to a perturbance for the structural system that may increase the load effect, influence the damage, and eventually increase the potential for failure leading to undesired consequences. The exposure consists of identifying the inventory exposed to perturbances at any point during the structure's lifetime. Finally, the vulnerability describes how the hazard may affect the structural system. The structural design is the primary driver of the vulnerability of the structural system.

A comprehensive risk assessment has qualitative and quantitative facets[18.13]. The qualitative dimension is oriented at properly describing the exposure, the hazards and the possible consequences. In this phase, the construction of scenarios is enough. Examples of tools adopted for the qualitative risk assessment are the hazard and operability study (HAZOP), fault tree, event tree, decision tree and causal networks. The quantitative dimension estimates the probabilities for all undesired events and associated consequences. This dimension allows the propagation of uncertainties with tools such as the probabilistic risk analysis (PRA). More about risk assessment is presented in Section 18.4. It is important to emphasise that the risk assessment can be repeated for multiple design solutions (or alternatives) to make the decision-making process more accessible.

18.3.3 Risk appraisal

Once risk is calculated for one design solution or multiple alternatives, this should be compared against the criteria defined in the risk management goal stage. This appraisal should be carried out as an optimisation exercise using cost-benefit-like analyses accounting for the multi-dimensionality of the problem (Figure 18.1).

An example of a risk appraisal approach is the use of the expected annual loss within a life cycle cost (LCC) or a return on investment (ROI) analysis to understand the effect of risk on the structure's life cycle from a financial feasibility and economic performance point of view, respectively[18.14].

The risk appraisal phase will have a clear homogeneous and consistent comparison of the different alternative design solutions as an output.

18.3.4 Decision making

The risk analysis is completed when it is possible to accept one alternative based on the risk appraisal outcome. However, in general, the stakeholder can have multiple decisions: (i) risk avoidance, (ii) risk reduction, (iii) risk transfer, (iv) risk retention. Risk avoidance means deciding not to proceed with the design if it is considered too risky. Risk reduction measures are design variations/revisions that are likely to generate a new risk that should be appropriately estimated. Risk transfer consists of transferring the risk to third-party organisations such as insurance companies; however, in this case, although the risk is shared, it does not change for the whole society. Finally, the risk is retained when it is deemed to be within acceptable levels according to the original risk management goals.

Examples of risk mitigation measures are:

- Hazard reduction/control providing appropriate structural and non-structural countermeasures
- Increasing the structural redundancy and resistance to degradation by improving strength and robustness
- Designing sacrificial elements (that can dissipate energy according to capacity design) where the probability of having consequences in terms of injuries or casualties must be minimal

Decision-making should be based on a mixture of engineering judgement and quantitative assessment. Both are necessary to ensure the decision-making process is robust.

18.3.5 Monitoring and review

The risk should be monitored for any of the four possible decision outcomes of the decision-making step to ensure it is kept within the desired target level. Monitoring will also guarantee continuous improvement of the process. A new risk analysis will be needed if the computed risk is no longer acceptable at any point in the structure's life cycle.

18.4 RISK ASSESSMENT

Risk assessment consists of the convolution of hazard, exposure and vulnerability (Figure 18.3). Such convolution is a non-trivial operation as uncertainties should be correctly propagated to the result after being inventoried and modelled. It is important to clarify that uncertainties are present in all three risk components; adopting deterministic approaches for any of these components implies a

Figure 18.3: General framework for structural risk assessment

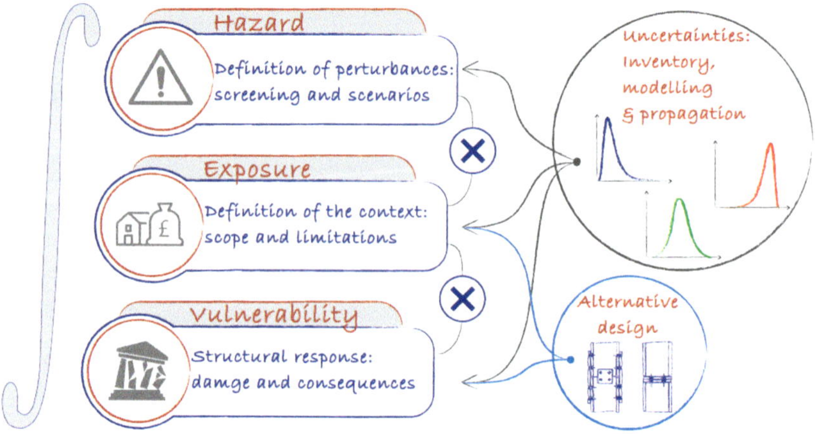

simplification of the assessment. Also, suppose different alternative design options are available. In that case, these need to be implemented at this stage to compute the risk for all of them, to simplify the comparative steps of the general risk analysis.

It is very convenient to decouple hazard, exposure and vulnerability in assessing risk. Specifically, according to the performance-based engineering framework, it is possible to think about a modular hierarchical and nested framework of components that can be organised according to the total probability law[18.15]. Equation 18.1 shows a simplified formulation for risk assessment[18.16] that is consistent with ISO 2394:

$$P\left(L \geq l\right) = \sum_{j=1}^{N_{DS}} \sum_{i=i}^{N_{IM}} P\left(L \geq l \middle| ds_j\right) \cdot p_{DS|IM}\left(ds_j \middle| im_i\right) \cdot p_{IM}\left(im_i\right) \qquad 18.1$$

where $P(L \geq l)$ is the loss function (or loss curve) representing the probability that the loss L exceeds a specific threshold l. The variables IM and DS are the hazard intensity measure and the structural damage state (see Section 6.2 for a definition of DS), respectively. The term p_{IM} (im) represents the probability mass function of the IM. The hazard analysis computing p_{IM} (im) is beyond the scope of this book; however, Chapter 10 provides some basic concepts on how to deal with such distribution. The term $p_{DS|IM}$ $(ds|im)$ describes the probability of attaining a specific damage state for any given intensity measure. Finally, the term $P(L \geq l|ds)$ is a conditional loss function obtained by combining a damage-loss function and the system cost model. The double summation is done to consider all possible discrete intensity measures and damage states.

A further simplification of Equation 18.1 is obtained by substituting the term $P(L \geq l|ds)$ with the expected loss associated with the specific damage state directly:

$$E\left[L\right]= \sum_{j=1}^{N_{DS}}\sum_{i=i}^{N_{IM}} E\left[L\left(ds_j\right)\right]\cdot p_{DS|IM}\left(ds_j \middle| im_i\right)\cdot p_{IM}\left(im_i\right) \qquad 18.2$$

where $E[L]$ is the expected (scalar) loss associated with the design at hand. It is essential to clarify that an accurate description of the perturbances always comes with a time dimension. If, for example, only the annual hazard maximum intensity measures are of interest, $p_{IM}(im_i)$ can be computed as an extreme value distribution (e.g., a Gumbel distribution — Section 10.3.2), and consequently, $E[L]$ refers to a single year (expected annual loss, EAL). The general method of changing the time horizon of the loss estimation starting from Equation 18.1 is:

$$P_{T=t}(L \geq l) = 1 - [1 - P(L \geq l)]^t \qquad 18.3$$

where t is the number of years and $P(L \geq l)$ is defined in Equation 18.1. When losses are quantified for a different time horizon, these should be discounted with an appropriate choice of discount rate (ζ). Specifically, the expected losses quantified in Equation 18.2 should be multiplied by the following factor:

$$\frac{1-e^{-\zeta t}}{\zeta} \qquad 18.4$$

A different approach can be followed if the hazard is provided in terms of the mean annual rate of occurrence $\lambda(im)$, or frequency-intensity curve. The rate concept can be introduced in Equation 18.1 as follows[18.17]:

$$\lambda\left(L\geq l\right)= \sum_{j=1}^{N_{DS}}\sum_{i=i}^{N_{IM}} P\left(L\geq l \middle| ds_j\right)\cdot p_{DS|IM}\left(ds_j \middle| im_i\right)\cdot\left|\mathrm{d}\lambda\left(im_i\right)\right| \qquad 18.5$$

where $|d\lambda(im_i)|$ is the absolute value of the differential of the mean annual frequency of exceeding the im_i. The term $P(L \geq l)$ on the left-hand side of Equation 18.1 is now substituted with $\lambda(L \geq l)$ that is the mean annual rate of exceedance of a specific loss value. In this case, the EAL can be derived in a simplified manner by substituting, in Equation 18.5, the term $P(L \geq l|ds)$ with the expected loss associated with the specific damage state directly:

$$EAL = \sum_{j=1}^{N_{DS}} \sum_{i=i}^{N_{IM}} E\left[L\left(ds_j\right)\right] \cdot p_{DS|IM}\left(ds_j \middle| im_i\right) \cdot \left|\mathrm{d}\lambda\left(im_i\right)\right| \quad 18.6$$

The EAL presented in Equation 18.6 corresponds to the area under the loss curve presented in Equation 18.5. Also, assuming a classical Poissonian arrival time process for the hazard, the probability of having a loss larger than a specific value in a specific time interval t will be:

$$P(L \geq l) = 1 - exp[-\lambda(L \geq l) \cdot t] \quad 18.7$$

In Equation 18.7, two main concepts can be emphasised. First, the inverse of the mean annual rate is equal to the return period (T_R). Second, for the specific case of a Poissonian arrival time process for the hazard, it is clear that the losses can be defined either using the mean annual rate of exceedance/occurrence $\lambda(L \geq l)$ or through the probability of exceedance/occurrence $P(L \geq l)$ and the reference time interval t. Building codes and national annexes often describe the actions in either of these two ways. All previous equations refer to a single hazard. Such analysis should be repeated for all the hazards of interest[18.18]. Eventually, losses associated with each hazard should be summed up. Correlation and causalities should be considered if cascading multi-hazard cases are considered. It is essential to clarify that the metrics used for losses should be consistent for such a summation. It is finally necessary to explain that the losses studied with the equations presented so far can be of any type: economic, human, environmental, etc. An example of this is given in Chapter 19 where carbon is considered in the case study. Therefore, the computational framework is common to all possible decision criteria. The issue of harmonising and homogenising the different calculated losses remains. Therefore, the risk acceptability becomes a multi-criteria optimisation.

18.4.1 Hazard

All actions, sources or situations that may cause undesirable consequences to the structure during its service life should be foreseen and characterised. This assessment step generally starts with the hazard screening phase. Specifically, hazards are chosen according to their significance with respect to the analysed structural system and in accordance with the nature of the site where the structural system will be built. Obviously, in this step, past experience and expertise of the designer will play a key role. The hazard is generally represented by a hazard curve mapping the intensity of a specific perturbance (e.g., wind gust) with its mean annual frequency (or alternative probability for a given observation time). The hazard phenomena are generally described using an intensity measure (IM), which is a scalar or a vectorial measure of the action (e.g., wind velocity, peak ground acceleration, axial loading). The IMs are estimators of the hazard phenomenon that is very complicated to describe; therefore, they should be robust, consistent, invariant, efficient and sufficient. All these properties are essential to ensure the accuracy of the structural performance assessment later on in the vulnerability model. The IM is generally a continuous variable (Figure 18.4a,c); however, it is very convenient to discretise the variable and consider only a number of hazard scenarios consisting of specific values of IM and corresponding probabilities/frequency (Figure 18.4b,c). This means passing from a probability density function

(PDF) to a probability mass function (PMF). This will simplify the convolution of the hazard component as the rigorous integral will become a summation, as shown in Equation 18.1. The discretisation of the IM allows considering a discrete number of cases, or scenarios, characterised by the discrete value of IM defining the specific class of events.

Figure 18.4: Discretisation of the hazard variable

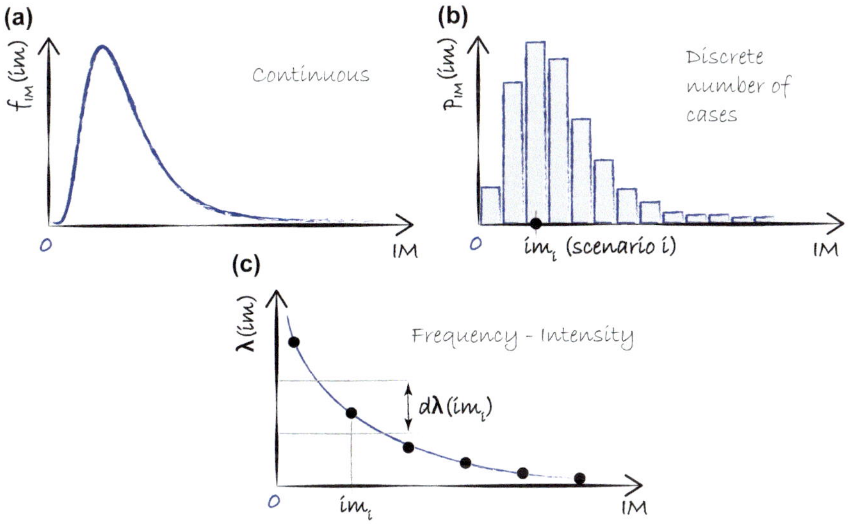

18.4.2 Exposure and impact

A multi-disciplinary estimate (e.g., survey engineering, actuarial math) of the exposed asset is needed. As clearly set out in *ISO 13824*[18.1], the task is to identify the spatial and temporal boundaries of the system involving the structural design at hand. A proper exposure assessment consists of identifying the economic, environmental and societal values of the asset that the possible structural failure mechanisms may influence. Such values are associated with the function of the exposed asset and should be assessed considering both the spatial dimension (e.g., structure footprint, neighbourhood, region) and the time dimension (e.g., service lifetime, centuries). Engineers may not be able to assess all these components; further competencies are needed for aspects related to societal values, for example.

Simple exposure quantifications are the total construction cost, the number of people during maximum occupancy and the tonnage of greenhouse gases produced to repair or rebuild. More complicated estimates are those associated with economic downtime due to the shutdown of the productive functions related to the structure (i.e., business interruption). The business interruption is not necessarily only related to the complete structural failure; it may also be due to the repair of a single component requiring the organisation of a building site.

Exposure assessment is not only quantification of the initial values but also an assessment of the consequences, i.e., an assessment of the losses when a given damage state is experienced. The classic tool developed for such a scope is known as the consequence function $f_{L|DS}(l|ds)$ (Figure 18.5a). These functions can be derived either from experience (e.g., historical structural failure) or can be computed for the design at hand with an actuarial exercise considering all possible damage states that the structure can experience (an example is provided in Chapter 19). The central value of the distribution is often provided as the product of a damage ratio and the total value of the asset. This central value is often used as $E[L(ds)]$ in Equations 18.2 and 18.6 to derive the expected losses quickly for a given scenario. The CCDF derived from the consequence function (Figure 18.5b) is the term $P(L \geq l|ds)$ in Equation 18.1. In the absence of information (i.e., if it is difficult to assess the losses for each damage state), uniform distributions are used for each damage state. Such uniform distributions are for damage ratios that are mutually exclusive. For example, let's consider a purely illustrative case (not corresponding to a real application) of three damage states: DS1, DS2 and DS3. To be mutually exclusive, the damage ratios for DS1, DS2 and DS3 must range between 0%–20%, 20%–80%, and 80%–100%, respectively. Sometimes the maximum damage ratio can exceed 100% because decommissioning, demolition and disposal costs are considered.

When the time dimension is considered, it is important to factor in all the discount rates for the actualisation of the values to a specified set of reference time intervals. Also, when casualties are computed, injuries can be factored in, relating them to fatalities. For example, homogenisation factors of 0.1 and 0.5 can be used to homogenise injuries and fatalities for moderate and major injuries, respectively. Discount effects on carbon emissions and other environmental impacts are not yet available and more research in this area is required, however, time-effects should be considered in the future.

The exposure stage also consists of assessing the cost of potential risk mitigation. Specifically, if different alternative design options are available, all of these need a proper holistic costing from a social, environmental and economic point of view.

Figure 18.5: Consequence functions for three damage states

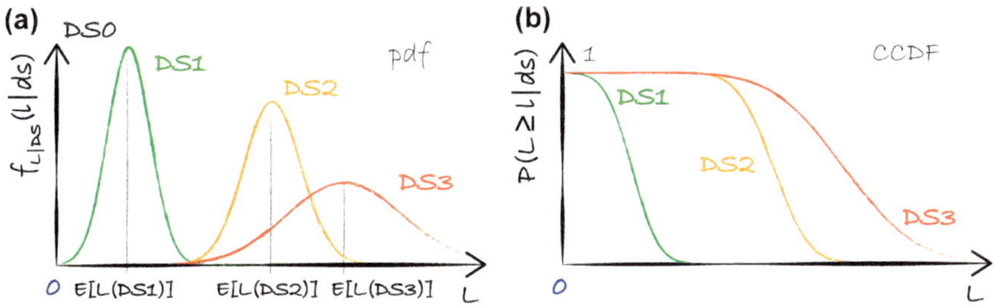

18.4.3 Vulnerability

Structural vulnerability is the susceptibility of a structural system to damage when subjected to an external action. In other words, vulnerability is the quantification of the ability/inability of a system to resist/respond to an adverse action[18.19]. From an analytical point of view, the vulnerability can be seen as a probabilistic relationship mapping a hazard intensity measure to a damage state and, in theory, the vulnerability should be hazard agnostic. This means that when a structure is subjected to a specific loading, this should not have peculiar characteristics that may not be representative of all possible situations. However, from a practical point of view, it is very convenient to have an initial idea of how to model the actions associated with the hazard. Chapter 6 presented the vulnerability as the simple comparison of load effects induced by external actions and structural capacity. In Equations 18.1 and 18.2, the structural vulnerability is represented by the term $p_{DS|IM}\ (ds_i|im)$ that is the probability that a given damage state is attained for a specific value of the hazard intensity measure. Specifically, the probability term $p_{DS|IM}\ (ds_i|im)$, which is a PDF, can be computed as the difference between the two probabilities of failure (CDFs in Equation A.5) conditioned on a specific level of hazard IM, associated with two consecutive limit states:

$$p_{DS|IM}\ (ds_i|im) = P_F\ (LS_i|im) - P_F\ (LS_{i+1}|im) \qquad 18.8$$

where the index i increases passing from light to severe limit states. For example, in Equation 18.3, if LS_i is the serviceability limit state, LS_{i+1} will be the ultimate limit state. It is also important to emphasise that in the summations presented in Equations 18.1, 18.2 and 18.3, the probability term $P_F\ (LS_{N_{DS}}|im)$ is equal to 0. The terms $P_F(LS_i|im)$ can be calculated with Equation A.5. All uncertainties can still be propagated in assessing the probability of failure, but the hazard intensity measure should be fixed to a specific value. Such a particular *im* value will lead to a specific set of load effect variables that should be compared with the capacities, as explained extensively in Appendix A.

As already mentioned in Chapter 6, the estimation of the probability of failure can be at a single component (e.g., column, beam) or for an entire structure with a systemic approach. If the vulnerability assessment is conducted element-by-element, the vulnerability model, and consequently the risk assessment, is known as component-based. Also, approaches exist to account for progressive collapse. These approaches, which allow the assessment of structural robustness, can be found in *ISO 2394*.

To avoid forgetting any correlation between damage (or failure) and consequence, it may be convenient to create a fault tree or a logic tree of the events (Figure 18.6). With such a framework, each damage state can be associated with a predefined consequence quantified with the exposure model. This approach also helps to describe precisely what a damage state is for the designer. However, it is important always to keep prescriptive aspects in mind. In fact, damage states are linked to the limit state definitions that are distinctly described in the building codes that should remain a reference for the designer. With reference to the schematic representation shown in Figure 18.6, it is possible to observe that for a specific value of intensity measure, the structure can experience no damage, or can reach either the serviceability limit state or the ultimate limit state. For the serviceability limit state, a potential expansion of damage assessment is provided, keeping in mind the alternatives that are provided in the building code. Each damage can be correlated to a specific consequence in terms of functional, economic, environmental and societal losses. The potential expansion for the ultimate limit state is left to the reader as a potential exercise.

Figure 18.6: Logic tree for correlating damage to consequences

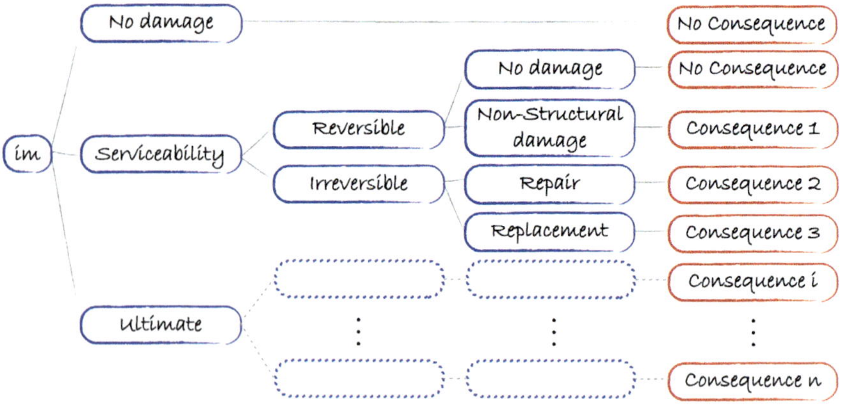

18.5 SIMULATION-BASED APPROACH

The general summation presented in Equation 18.1 can be solved using a simple simulation-based procedure. Such a procedure becomes very useful when many uncertainties are going to be considered and propagated in the final result. Figure 18.7 shows two possible approaches: a complete simulation-based approach (Figure 18.7a) and a partial simulation-based approach (Figure 18.7b).

Regarding the complete simulation-based approach, the first step is the simulation of a field of intensity measures ($\mathbf{im} = [im_1, im_2, \ldots, im_K]$, with K being the number of actions — or hazards — acting on the structure) using the probability distributions describing the perturbances $f_{IM,K}(im)$. The second step consists of computing the probability of attaining a specific damage state $p_{DS|IM}(ds_i|im)$ for each simulated vector of intensity measures. For this step, instead of computing the probability, it is possible to assess whether a limit state is exceeded for each simulation. In doing so, all the epistemic (i.e., lack of knowledge) uncertainties can be simulated according to what is explained in Chapter 6. At the end of this step, it will be possible to identify which damage state is obtained for the specific scenario (i.e., realisation) by comparing load effect (demand) and capacity. It is important to clarify that load effect and capacity may be dependent on the simulated hazard and epistemic variables; this is depicted with dashed arrow lines in Figure 18.7. According to the achieved damage state it will be possible to assess and quantify the consequences using Figure 18.6. A single value of consequence (e.g., economic loss, CO_2 emissions) can be extracted using the distribution presented in Figure 18.5. By repeating the iterations many times (according to Equation 6.9), it will be possible to have a large number of simulated consequences that can be used to approximate Equation 18.1 or compute Equation 18.2.

Figure 18.7: Simulation-based approaches for computing risk

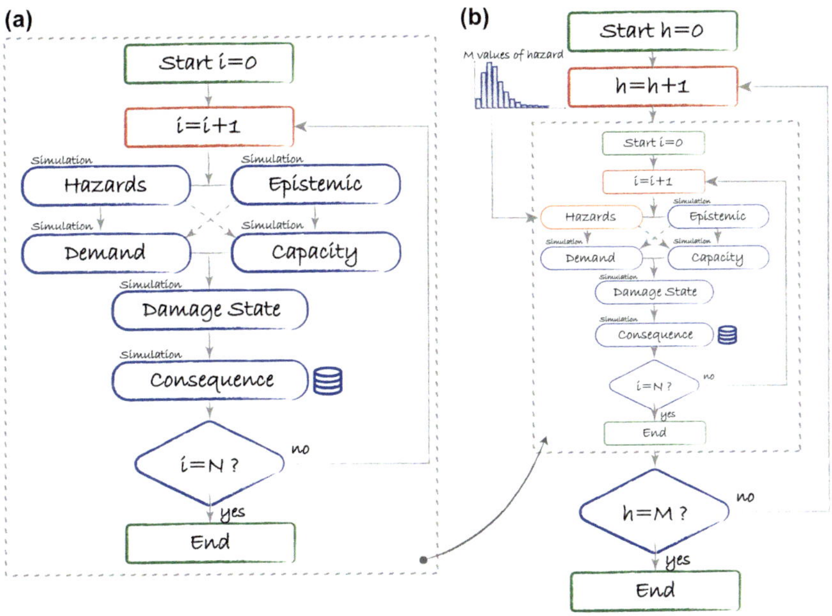

Alternatively, a partial simulation-based approach can be used (Figure 18.7b). This approach consists of avoiding the

simulation of the hazard components (usually the most important regarding the level of overall uncertainty) using the discretised distribution of the hazard and repeating the simulation procedure on all the other components (Figure 18.7a) for a number of deterministic values of IMs. Eventually, Equation 18.1 can be used to combine everything. This reduced simulation scheme can be helpful for a small number of perturbances and a first risk screening.

It is interesting to observe that, with reference to Figure 18.7a, if the simulation procedure is stopped at the damage state assessment, then the analysis is a simple reliability-based analysis. It is necessary to move to the consequences in order to have a proper risk-based approach.

18.6 RISK REPORTING

Risk analysis results should be reported with utmost clarity at every step of the process. Qualitative risk reporting is generally related to adjectives such as high, moderate and low. In quantitative terms, risk should be presented by relating consequences to probability. Both qualitative and quantitative representations should include plenty of information to allow decision making. As far as structural engineering is concerned, quantitative reporting should be preferred. Some risk representations are provided in Sections 18.6.1 and 18.6.2.

18.6.1 Loss curve

The loss curve is a representation of the complete profile of the risk. Figures 18.8a and 18.8b represent the loss curves obtained with Equations 18.1 and 18.3, respectively. Such curves can be represented in the linear space (Figure 18.8a) and the log-log space (Figure 18.8b). The representation in the log-log space is often preferred as it provides better visualisation of the data for both rare and frequent events.

Such curves give the possibility of a graphical comparison between the sensitivity of the design to different hazards (Figure 18.8a) and the effect of different design alternatives (Figure 18.8b).

For example, Figure 18.8a shows that *Hazard 1* is more important for more probable events, and *Hazard 2* becomes dominant for rare events. Similarly, Figure 18.8b shows that the design *Alternative 1* returns lower losses for more frequent events (i.e., high values of λ) compared to the design *Alternative 2*. However, *Alternative 1* leads to disproportionate losses for rare events (i.e., low values of λ) compared to *Alternative 2*. It is important to clarify that for representation associated with specific scenarios, the maximum value expected for the exceedance probability is 1 (Figure 18.8a). Conversely, a loss curve represented in terms of mean annual rate (Figure 18.8b) can exceed 1.

If the loss is quantified in terms of casualties, such curves are also known as F–N curves or frequency–number (of casualties) curves. Such F–N curves are used as a simple measure of societal risk, and they are often represented in the form αN^{-K} where α and k are two parameters. The cases $k < 1$ and $k > 1$ represent risk processes and risk aversion cases; $k = 1$ stands for risk-neutral cases. For acceptable risk in case of structural failure, an acceptable value of α is 10^{-6} [18.9].

Figure 18.8: Loss curves for (a) probability and (b) mean annual rate

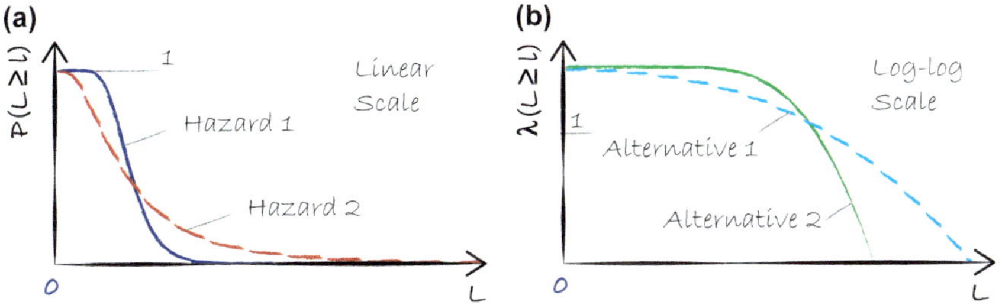

In absolute terms, to make sure that the computed risk is acceptable, the risk curves can be compared with some limits (Figure 18.9). The risk should be contained as far as possible in the acceptable and ALARP region. For example, a cost-benefit analysis can be carried out to decide if the risk is tolerable in the ALARP region. With respect to the ALARP framework, the upper limit is often easily defined. The lower limit is more flexible. Finally, such a framework can be used to assess the risk from all perspectives, i.e., societal, environmental and economic.

Figure 18.9: General ALARP principle

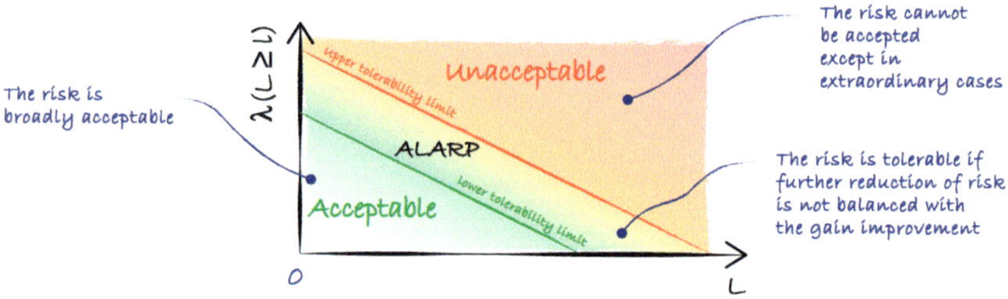

18.6.2 Risk matrix

The arrangement of losses and probabilities in a matrix (Figure 18.10) can be a suitable risk representation for qualitative or semi-quantitative estimations at the early design stages. Also, in this case, the matrix is separated into acceptable, ALARP and unacceptable risk regions. In the ALARP region, it is necessary to investigate whether further risk reduction is needed or if more detailed studies should be conducted.

The probability of exceedance can be ordered from very low (rare events) to very high (frequent events). The loss axis can be discretised in three or more classes. For the case of three classes, it is possible to describe them as (a) small, (b) medium and (c) catastrophic. In the case of five classes, they can be: (a) very low, corresponding to local damage of small importance, (b) low, corresponding to local damage, (c) medium, corresponding to the failure of a part of the structure leading to a small chance of injuries, (d) high, corresponding to a partial failure of the structure with a possibility for partial collapse and several injuries and (e) severe, corresponding to a sudden collapse of the structure leading to potential loss of life and several injuries.

Figure 18.10: A risk matrix

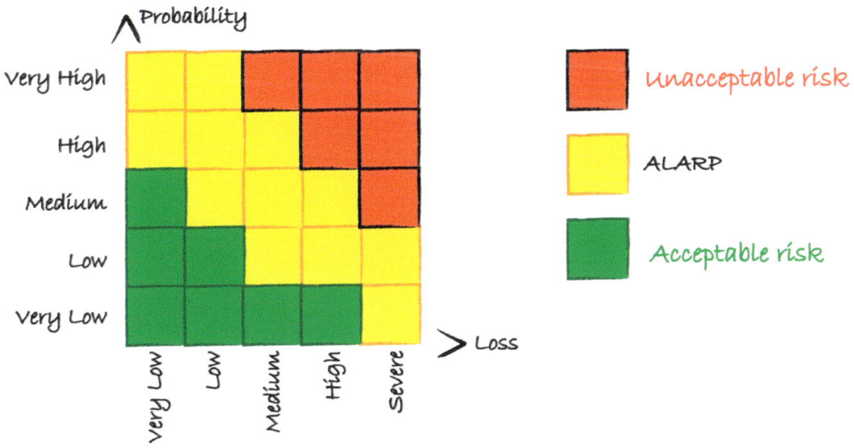

18.7 CONCLUSIONS

In this chapter, we have taken a concept that engineers understand well: the assessment of risk. We have also shown how a quantitative risk-based design can be carried out. In Chapter 19, we will return to the case study from the end of Parts 2 and 3 and consider what this might look like if we apply a risk-based design approach, and how we can consider both financial and carbon accounting in our decision-making process.

REFERENCES

18.1 *ISO 13824:2020: Bases for Design of Structures — General Principles on Risk Assessment of Systems Involving Structures.* Geneva: ISO, 2020.

18.2 *ISO 2394:2015: General Principles on Reliability for Structures.* Geneva: ISO, 2015.

18.3 UK Structural Engineers Declare Climate and Biodiversity Emergency. Available at: https://www.structuralengineersdeclare.com/ [Accessed: January 2025].

18.4 Melchers, R.E. and Beck, A.T. *Structural Reliability Analysis and Prediction*. Hoboken: John Wiley & Sons, 2017.

18.5 Schneider, J. 'Safety — A Matter of Risk, Cost and Consensus'. *Structural Engineering International*, 10(4), November 2000, pp266–269. Available at: tandfonline.com/doi/abs/10.2749/101686600780481347 [Accessed: January 2025].

18.6 HM Government. *Reclaiming Health and Safety for All: An Independent Review of Health and Safety Legislation (Löfstedt Report)*. 2011. Available at: https://www.gov.uk/government/publications/reclaiming-health-and-safety-for-all-lofstedt-report [Accessed: January 2025].

18.7 Rackwitz, R. 'Optimization and Risk Acceptability Based on the Life Quality Index'. *Structural Safety*, 24(2–4), April–October 2002, pp297–331. Available at: https://doi.org/10.1016/S0167-4730(02)00029-2 [Accessed: January 2025].

18.8 The Institution of Structural Engineers. *How to calculate embodied carbon (Third edition)*. London: IStructE Ltd, 2025.

18.9 Diamantidis, D. 'A Critical View on Environmental and Human Risk Acceptance Criteria'. *International Journal of Environmental Science and Development*, 8(1), January 2017, pp62–66. Available at: https://doi.org/10.18178/ijesd.2017.8.1.921 [Accessed: January 2025].

18.10 *NORSOK Z-013:2024: Risk and Emergency Preparedness Analysis*. Lysaker: Norway, 2024.

18.11 UNDRR. *Technical Guidance on Comprehensive Risk Assessment and Planning in the Context of Climate Change*. Geneva: United Nations Office for Disaster Risk Reduction, 2022. Available at: https://www.undrr.org/publication/technical-guidance-comprehensive-risk-assessment-and-planning-context-climate-change# [Accessed: January 2025].

18.12 The Institution of Structural Engineers. *Conceptual design of buildings*. London: IStructE Ltd, 2020.
18.13 British Standards Institution. *BS EN 1991-1-7:2006+A1:2014: Eurocode 1. Actions on structures - General actions - Accidental actions*. London: BSI, 2014.
18.14 De Risi, R. *et al.* 'Life Cycle Cost and Return on Investment as Complementary Decision Variables for Urban Flood Risk Management in Developing Countries'. *International Journal of Disaster Risk Reduction*, 28, June 2018, pp88–106. Available at: https://doi.org/10.1016/j.ijdrr.2018.02.026 [Accessed: January 2025].
18.15 Jaynes, E.T. *Probability Theory: The Logic of Science*. Cambridge: Cambridge University Press, 2003.
18.16 Ciampoli, M. *et al.* 'Performance-based Wind Engineering: Towards a General Procedure'. *Structural Safety*, 33(6), September 2011, pp367–378. Available at: https://doi.org/10.1016/j.strusafe.2011.07.001 [Accessed: January 2025].
18.17 Sevieri, G. and Galasso, C. 'Typhoon Risk and Climate-change Impact Assessment for Cultural Heritage Asset Roofs'. *Structural Safety*, 91, July 2021. Available at: https://doi.org/10.1016/j.strusafe.2020.102065 [Accessed: January 2025].
18.18 Venanzi, I. *et al.* 'Multi-hazard Loss Analysis of Tall Buildings Under Wind and Seismic Loads'. *Structure and Infrastructure Engineering*, 14(10), February 2018, pp1295–1311. Available at: https://doi.org/10.1080/15732479.2018.1442482 [Accessed: January 2025].
18.19 Agarwal, J. *et al.* Vulnerability Analysis of Structures. *Structural Engineering International*, 16(2), May 2006, pp124–128. Available at: https://doi.org/10.2749/101686606777962611 [Accessed: January 2025].

CHAPTER 19

Future case study — Part 3

Raffaele De Risi

19.1 INTRODUCTION

For a third time, we return to the case study of our single-story rammed earth house with timber joists. In this final part of the case study, we explore how we might consider using a risk-based design approach. The case study also explores incorporating both cost and embodied carbon into the calculation. As with previous parts of this case study, this is not a 'how-to' but instead intends to explore the possibilities that this approach might enable.

19.2 CASE STUDY — ADDITIONAL KEY INFORMATION

In Section 9.2 we defined the geometry, materials, limit states, capacity, load effect and loading. We now need to define exposure.

19.2.1 Exposure

As explained in Chapter 18, the exposure component consists of defining the value of the asset at risk and the potential consequences of the damage to the system involving the structure. For simplicity and clarity, in this example, two exposure aspects have been considered: economic and environmental dimensions. The building contents and the potential casualties/injuries are not part of this case study, however, they can be considered similarly, as well as the indirect costs due to business interruption and downtime in general.

From an economic aspect, the 2020 Spon's Architects' and Builders' Price Book[19.1] was adopted. In case of exceedance of the SLS, an average repair cost of £80/m^2 was considered. For the ULS, an average price of £225/m^2 was considered for the roof components and joists. A cross-section-dependent value was used for the latter from the price book. For the ULS, in addition to the value of the structure, a further cost of about £35/m^2 was added for demolishing and disposal. This number is approximated as it also depends on the dimensions of the joists. In the worked example, this number is computed rigorously for different design solutions.

From an environmental aspect, CO_2 emissions were computed according to *A brief guide to calculating embodied carbon*[19.2]. The CO_2 emissions are calculated precisely according to the typology of construction activity and the mass of used material. The asset lifetime is considered to be 50 years (although we would hope, with good maintenance and a periodic structural investigation, this could be extended significantly), and the component lifetime is assumed to be 20 years. Also, the transportation to site is assumed as a random variable distributed according to a normal distribution with a mean equal to 200km and a standard deviation equal to 50km. Finally, the amount of wasted material is also simulated according to a uniform distribution between 0% and 100%; this is because the amount of wasted material cannot be measured precisely.

For both the economic and environmental components of the exposure model, a lognormal distribution is adopted for describing the consequence distribution $f_{L|DS}(l|ds)$ (Chapter 10). Specifically, the mean value is assumed to be equal to the nominal value, and the variation coefficient is assumed to be 10% for the economic consequences and 30% for the environmental consequences. This assumption is made because there is more uncertainty associated with the quantification of CO_2 emissions.

19.3 RISK-BASED ANALYSIS OF TIMBER JOIST

Following the steps laid out in Sections 19.3.1 and 19.3.2, we will now analyse the timber joist, which was originally analysed using reliability methods in Chapter 9.

19.3.1 Risk management goals

The first step is the definition of risk management goals. For the considered case study, some simplifications can be defined. Specifically, risk goals are defined in terms of risk acceptability for the two considered limit states, i.e., serviceability and ultimate limit states. For the SLS, a loss equal to the associated expected value of the consequence function (i.e., a repair cost of £80/m^2 and a CO_2 emission corresponding to the repair operations) can be assumed to be acceptable either for a return period of 50 years or for a probability of exceedance corresponding to the reliability index of 2.9. Analogously, for the ULS, a loss equal to the associated expected value of the consequence function is assumed to be acceptable for a probability of exceedance corresponding to a reliability index of 4.7.

Figure 19.1 shows the risk target in terms of ALARP upper tolerability limit; losses are represented for the system concerning a single joist. Specifically, the example of a C14 joist having cross-section 75 × 200mm and 300mm spacing is considered. The figure shows two different options of functional forms and the two different options associated with the acceptability of the SLS. For the SLS, adopting a reliability index of 2.9 rather than the return period of 50 years will provide a more conservative approach to the risk appraisal phase. A more conservative approach is obtained if a linear functional form in the semi-logarithmic space is adopted rather than a linear form in the linear space.

Figure 19.1: Example of risk goals for both (a) economic and (b) environmental losses

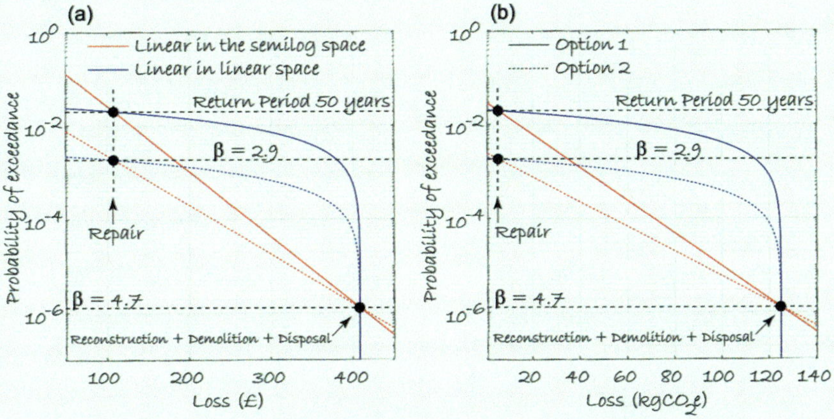

The linear functional form in the semi-logarithmic space is adopted for this case study. However, the SLS consequences are assumed to be acceptable for a return period of 50 years. Also, as the upper tolerability limit changes for each design solution, the design options cannot all be represented on the same plots. Multiple plots are needed for the sake of clarity. Table 19.1 lists the expected annual losses, and Figure 19.2 shows the loss curves computed for different joist cross-sections. It is possible to conclude that the cross-section 75 × 200 is the minimum section satisfying the risk goals and, therefore, is acceptable according to the risk appraisal phase. As shown in Figure 19.2, it could be informative to plot (on top of the loss curves) the cases in which the ULS is exceeded. These are marked with an additional yellow point; all other points on the plotted lines represent SLS exceeded. This will provide further information to communicate and discuss with the decision-makers. For example, it is possible to argue that for the cross-section 75 × 200 the ULS cases start appearing for a probability value that is acceptable according to the discussed values of reliability. For the cross-section 75 × 250, the probability of exceedance is so tiny that it cannot be visualised with the adopted axes limits.

It is interesting to observe that the loss curve has two main parts; this feature reflects the fact that the two consequence functions associated with the two limit states are quite distinct and well separated.

Figure 19.2: Single joist loss curves in (a–d) economic and (e–h) environmental terms for a joist spacing of 300mm and for the cross-sections (a, e) 75 × 175, (b, f) 75 × 200, (c, g) 75 × 225 and (d, h) 75 × 250

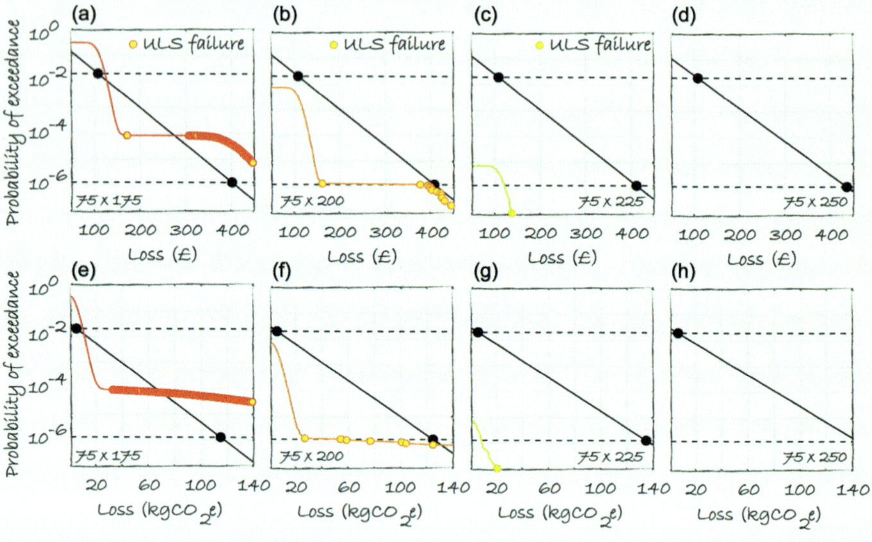

Table 19.1: Expected annual losses for a single joist

	75 × 175	75 × 200	75 × 225	75 × 250
£	32.9	0.77	$0.75 \cdot 10^{-3}$	0
$KgCO_2e$	1.64	0.04	$0.06 \cdot 10^{-3}$	0

Having completed the analysis it is possible to postulate that the overall carbon could be calculated by considering the embodied carbon at the point of construction and adding the future additional carbon due to repair for SLS and ULS cases. This approach is tentatively presented in Table 19.2. The outcome is only tentatively presented as it is a future postulation of what might be; it still contains much uncertainty. The table is constructed by taking the assumed breadth, depth and length to find the section volume. This is then multiplied by the mean density to give a mass. The mass of the timber is then multiplied by the embodied carbon factor (ECF) taken from Table 2.5 in *How to calculate embodied carbon*[19.3]. This gives an embodied carbon for the joist at the point of construction (cradle to gate). We then take the expected annual loss, as presented in Table 19.1, and combine them, to show that the embodied carbon is optimum in the 75 × 200 section. This is a highly simplified approach and does not account for discounting of embodied carbon over time, as you would for cost — an area which requires considerable research. However, it also suggests that there could be an optimum size, when considering embodied carbon from both construction and expected losses due to repair. This is an example of design optimisation as presented in Figure 18.1. Here we are minimising the carbon rather than maximising the benefit.

Table 19.2: Calculating and combining embodied CO_2e with expected annual loss

Breadth (mm)	Depth (mm)	Length (m)	Density (kg/m³)	Volume (m³)	Mass (kg)	ECF	Embodied at construction kg CO_2e	Expected annual losses kg CO2e	Embodied plus loss kg CO_2e
75	175	5	350	0.066	23.0	0.263	6.04	1.64	7.68
75	200	5	350	0.075	26.3	0.263	6.90	0.04	6.94
75	225	5	350	0.084	29.5	0.263	7.77	6.00E-05	7.77
75	250	5	350	0.094	32.8	0.263	8.63	0	8.63

19.3.2 Sensitivity analysis

When a risk-based design is conducted, it is always a good idea to perform a sensitivity analysis to identify which uncertainty affects the design the most[19.4]; this step is also beneficial for identifying any problems with the implementation. For this case study, it has been noted that the correlation between the mechanical properties, the demand variability and the model variability have a minimum effect on the final results, acting slightly only on the very low probability of exceedance. The uncertainties associated with the consequence function and the uncertainties on the geometry have the most significant effects. Specifically, the variability of the consequence function affects the shape of the loss curves, which become almost vertical, showing determinism in the potential consequences. The influence of the variability on the geometric properties is more significant. Figure 19.3 shows the case of considering and neglecting the uncertainties about the geometric features. Ignoring such uncertainties leads to a larger probability of exceedance of the losses and, therefore, a potential lack of optimisation in the design of the structure. This phenomenon, for this case study, is because the denominator of the limit state functions (i.e., the demand) on average increases by 20%, 15% and 10% when geometric uncertainties are neglected for the SLS, ULS ductile and ULS brittle, respectively. This is a consequence of how the cross-section properties are probabilistically modelled.

Specifically, when the uncertainties are considered, the cross-section dimensions are slightly larger due to the adopted approach (Section 9.2.1). Such an increase, in this chapter, is considered consistently for the assessment of the cross-section area, elastic modulus and inertia moment. These mechanical characteristics of the cross-section are larger if the geometric uncertainties are considered rather than neglected. Larger cross-section mechanical values

Figure 19.3: Loss curves in (a) economic and (b) environmental terms for a joist spacing of 300mm and the cross-sections 75 × 225

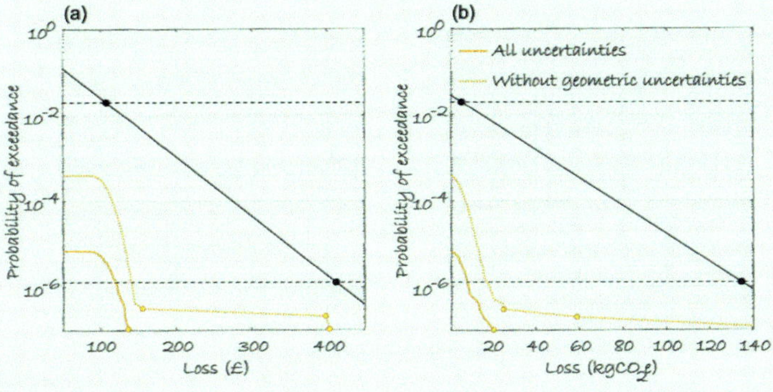

have a small impact on the self-weight and return lower stresses on the section on average and, therefore, smaller demand. Such an aspect can be considered an issue; therefore, it can be avoided using the uncertainties on the cross-section only for the loading model and not for definition of the section properties. In this latter case, when the geometric uncertainties are neglected, the loss curves are actually smaller than when uncertainties are considered.

This long discussion is to strengthen the importance of critical thinking about the results and the importance of sensitivity analysis.

19.4 CONCLUSIONS

In this final speculative case study we have started to explore the possibilities available to us when we apply risk-based design approaches. We have seen how a more complete and nuanced understanding of risk can be explored, how we can utilise different parameters to measure the preferred outcome within an ALARP context, how a sensitivity analysis can highlight possible areas of concern and the need for engineers to apply critical appraisal rather than assuming the computed outcome is correct — a reminder that putting rubbish in, gives rubbish out.

REFERENCES

19.1 AECOM. *Spon's Architects' and Builders' Price Book.* Abingdon: Routledge, 2019.

19.2 Orr, J. *et al.* 'A brief guide to calculating embodied carbon'. *The Structural Engineer*, 98(7), July 2020, p6. Available at: https://doi.org/10.56330/JZNX5709 [Accessed: January 2025].

19.3 The Institution of Structural Engineers. *How to calculate embodied carbon (Third edition).* London: IStructE Ltd, 2025.

19.4 Pianosi, F. et al. 'Sensitivity Analysis of Environmental Models'. *Environmental Modelling and Software*, 79, May 2016, pp214–232. Available at: https://doi.org/10.1016/j.envsoft.2016.02.008 [Accessed: January 2025].

CHAPTER 20

Epilogue

James Norman

20.1 INTRODUCTION

At the end of this book I want to break cover and be both honest and personal. I was tempted to write a future imagining, just as I did for the epilogue of *The regenerative structural engineer*[20.1]. But if you've read that book you will have already seen just how much of the thinking in this book is also in that publication. For example:

> "*We have high-end computers for advanced analysis. We no longer run simple linear analysis. Instead, we use enhanced processing power to run non-linear analysis to optimise design and if possible, we run a full probabilistic analysis. That way we can better understand how likely something is to fail. Where projects have a negative impact (in the form of loss of biodiversity, pollution, emissions, etc.) we run a detailed risk analysis to ensure that the risk of failure is optimised against the environmental risks.*"

Rather than repeat the whole chapter, I wanted to finish with an opinion piece; a reflection on what has gone before, and a challenge to think hard about what happens next.

This book is called *The future of structural design*, but it has had many working titles as it has gone through countless iterations and different imaginings. At one point it was over-optimistically called 'How to design absolutely anything'. At another it was dourly called 'Risk and reliability-based design'. The title changes in part match the struggle with the subject. Is it a probability textbook or a 'how to' or some future imagining? Does it just cover the numbers-side or incorporate the bigger picture of the design of materials, both to the code and beyond? What has remained, from when I first imagined the book back in 2012 to when I first realised it could actually be written with the right people in 2020, to this finished version is the desire to understand uncertainty better — both risk and reliability — so that we can do less harm and do more good.

'Do no harm' was the standard I, along with everyone else, was working to back in 2012. However, 'actively doing good' is how I have now come to think about it having written *The regenerative structural engineer*.

In writing this book we have tried to pull together different elements of ethics, testing, design approaches and probability theory, and write a book which captures something of the future of design. That future is unknowable. The content is just an imagining, a model, an approximation of what might be, just as our structural analysis is a model of what we imagine in the future. There is much uncertainty to contend with.

The same is true around the uncertainty of our species and many others[20.2]. We know that we have already crossed two of our planetary boundaries as many of us work for companies who have declared a biodiversity (now referred to as 'biosphere integrity') and climate emergency[20.3]. The reality is that we have crossed six of the nine planetary boundaries[20.4]. We know that our work has a huge impact on climate emergency, and

Buildings and construction currently account for around 40% of energy-related CO_2 emissions[20.5].

ways to quantify the impact our work has on the biodiversity emergency across the supply chain (not just on the site where we are building) are starting to emerge[20.6,20.7]. But what impact do our materials have across the other seven boundaries? Where are we responsible for specifying materials that impact on land–system change, freshwater change, biogeochemical flows and novel entities — the other four boundaries already crossed? The answer is we don't yet know, but we need to discover.

20.2 ...AND NO MORE

What we do know is that engineers use too much. And the first practical step that I hope you take from this book is to 'design to the code …and no more'. Of course, designing to the code could have many meanings, from a quick calculation using a number of simplifying assumptions to speed up the design, to making every effort to utilise every aspect of the code to ensure our design is as efficient as possible. In the past I have taken the first option, to make life simple and enable me to design as many beams and columns as possible as quickly as possible. But there is now (and always has been) an ethical imperative to do better. The imperative comes in two forms: the 'ability to pay' and 'one actor in a much larger, unfair system' (Section 3.3). These both affect us as designers, because we can afford to design better, and we do, as engineers, have agency to make these changes (although we also must manage the challenges of the larger, unfair system). These two also interplay as the fees that we charge are often what drive us to design quickly, and yet the fees are not often set by us, but by the market. The imperative also, and more importantly, comes from the fact that we now know that those who have done the least to contribute to climate change, those in the global south, will experience the greatest loss and impact[20.8,20.9].

As engineers we have a responsibility to take safety incredibly seriously. I have lain awake at night worrying about a design, and then double-checked it and triple-checked it in the morning to make sure it is safe. This is absolutely the right thing to do, but we should have the same sense of responsibility to make sure we have not over-designed.

This is an interpretation of *'Engineering professionals have a duty to obey all applicable laws and regulations and give due weight to facts, published standards and guidance and the wider public interest.'*

In fact, it is part of our professional commitment to protect the public[20.10], and to:

- Protect, and where possible improve, the quality of built and **natural environments**
- Maximise the public good and **minimise** both actual and potential adverse effects for their own and succeeding generations
- Take due account of the **limited availability** of natural resources

Not only are we ethically bound to do this, but we are also professionally bound.

This will in part come from better quality assurance. Which may feel like a strange leap, but quality assurance means that we can have confidence in our design. We have reduced the chance of human error by double-checking the design principles and assumptions we have made; by double-checking the calculations we have performed to make sure they are both safe and not conservative and ensuring what we have designed is what has been built. We can also utilise quality assurance in the construction process through more prefabrication and off-site construction, which is both safer and has the possibility to reduce uncertainty, in geometric terms and in terms of materials. This information can then be utilised, together with the codes, to make our design even more efficient.

If you find yourself lying awake at night worrying about the design, or if you add a bit of extra 'bunce' just to make sure — this is a sign that you are not confident that the design is right. The solution should not be to make it bigger, but to ask for help and check it, to use physical models to ensure the analysis really is predicting the behaviour we would expect. Bring in an expert if necessary. We have an ethical responsibility to design it to the code and no more. I understand this is a change in culture, but I also know that some people are already doing this. They should be shouting about it, and we should be celebrating them.

We should also be asking our institutions, code committees and governments if the balance is right. Have we got the local vs. global risk right? Codes are not static, they evolve.

One final thought before we move on — designing to code is just following a series of complex rules. In the not too distant future computers will be able to do this much better than we can (I am aware that they already can in a number of discreet areas). The role of the engineer is changing. Designing to the code and no more may well be taken out of our hands, but we still need to have a deep understanding, in fact it is even more important, not just of the main clauses, but all clauses, so that we can make sure they have been used correctly and with confidence. We need to ensure that the analysis is a good enough approximation of the world we plan to build. We may soon find that we don't just have an ethical driver but other reasons for being at the forefront of designing to code — utilising every opportunity to make our designs more materially efficient.

20.3 REGENERATIVE DESIGN

Of course, doing less harm is no longer good enough. We need to find a way to go beyond 'do no harm' and instead to 'actually do good'. We need to find ways to create materials that reduce greenhouse gases, increase biodiversity, do not create more novel entities, improve land use, provide more and better-managed freshwater while managing biogeochemical flows. Some of those solutions exist, in the form of known regenerative growing practices, others are still to be discovered. Rather than write in detail a treaty on regenerative design here, I encourage you to read any of the excellent books on the topic that currently exist[20.1,20.11–20.13].

An important point to note is that many, if not all, of the materials and approaches that are required by the regenerative structural engineer either don't yet exist, exist but are not presently codified, or are codified generally, but the classification is generic and does not cover the subtleties of local supply chains. In all three cases we first need a better holistic understanding of the materials and the way they behave. We need to understand what will govern the design and where the risk lies, and we need to quantify the uncertainty in these areas.

We also need better understanding of the uncertainty in the quantification of the impact across all nine planetary boundaries so we can make positive and informed decisions about our materials. As we apply a continuous, place-based design process, we need to generate the information that we can feed into our decision-making process to improve our decisions systematically. This requires us to share as much data as we can with each other, not just the good news stories, but all the information, and we need to avoid ethical disasters, such as that which beset the manufacturers of diesel cars over the last decade or so[20.14].

20.4 REUSE

I have, over the last few months, found myself standing on a hill or rooftop and looking out over a city, whether my home city of Bath or my work city of Bristol (Figure 20.1). As I look out over these cities I am struck by the breadth and depth of physical material wealth we have in the built environment in the global north. I find it hard, maybe impossible, to believe that in these cities we do not already have enough. It may not be in the right place or the right form but we are extraordinarily materially rich. I have for some time, alongside many others, been advocating for no more new buildings[20.15,20.16]. In writing this book, I am more than ever convinced that we have an ethical responsibility to make reuse, deconstruction and reconstruction the norm where we already have this material wealth.

Figure 20.1: View over Bath from Alexandra Park

This will not happen overnight, however — we need to change the system. Some of that system change has started. We highlight new codes which have recently been published in Part 4, and in the introduction we discuss IStructE's own book on circular design, *Circular economy and reuse: guidance for designers*[20.17]. What anyone actively involved in designing for reuse or with reused components will be able to tell you is that we don't yet have a consensus on how as an industry we should be doing this. Methods are emerging, and over the next few years I expect to see a lot more being published, but currently there is no 'how to' guide for designing for reuse that adequately covers uncertainty. Many practitioners have either:

- Looked to use the modern codes — but these are not designed for reuse, and much of the uncertainty around new buildings is no longer present (we can measure the size and out-of-plumb with much greater confidence) while new uncertainties arise (are there signs of degradation, water ingress, distress and how do we manage those?), or
- Returned to codes utilised at the time the building was designed and built

While both are helpful to a certain degree neither is adequate to check an existing building fully.

Again, by sharing information from the strength of materials tested to the levels of quality assurance and build quality as a function of when work was done, we can try to understand the problem better. There will always be uncertainties and we will need to manage them. However, as with the award-winning Elizabeth Line OLE gantry (Case Study 6 in *The regenerative structural engineer*[20.1]) and the case study in Chapter 17 of this book, in-depth observation of the existing structure, followed by rigorous analysis that looks to understand design beyond the limitations of our codes, can lead to substantial reuse of structures. Furthermore, as with the Elizabeth Line OLE gantry, through the sharing of both approaches and data, we can enable others to do the same.

20.5 THE FUTURE OF DESIGN

The future of design is uncertain.

It is both uncertain and will involve modelling uncertainty.

As our understanding of the role and impact of our work becomes clearer across all nine planetary boundaries, the moral mandate to do better rises, and our ability to manage and compute complexity grows. I believe we will need to, at times, reach beyond reliability and into full risk analysis.

This will be a huge step for practising engineers and should not be taken lightly.

It is essential that outcomes can be sense-checked and verified by an engineer. We will need designs that we know are safe and we can show through calculation are safe. We will also want to understand the uncertainty in the risk for all the different aspects we can control.

This is not a simple undertaking.

However, in the final case study (Chapter 19) we have a tiny glimpse of what might be. We start to look at the risk both financially and in carbon terms.

At present we have an acceptable risk of building failure. It is possible to imagine that in time we might have a way to quantify and define the climate risk, and the biodiversity risk, and novel entity risk for new materials, and these, too, could be quantified and controlled for our buildings.

We may then find ourselves, as the dam designers and nuclear power station designers today, considering the ALARP across these multifaceted risks to present a case for why the design is necessary and safe enough, both locally and globally.

The world is changing. The role of the engineer is shifting. We have an incredible opportunity in front of us to show ethical leadership in the built environment, raise the value of our occupation and do some great fun engineering, playing both with numbers and materials.

Some nights I lie awake worrying about the future …but in the daytime, as I think about all the ideas bursting from this book, I feel genuinely excited about the future of design and the role we will all have the opportunity to play in it.

REFERENCES

20.1 The Institution of Structural Engineers. *The regenerative structural engineer.* London: IStructE Ltd, 2024.

20.2 Kolbert, E. *The Sixth Extinction: An Unnatural History.* London: Bloomsbury Publishing, 2014.

20.3 UK Structural Engineers Declare Climate and Biodiversity Emergency. Available at: https://www.structuralengineersdeclare.com/ [Accessed: January 2025].

20.4 Stockholm Resilience Centre. *Planetary Boundaries.* Available at: https://www.stockholmresilience.org/research/planetary-boundaries.html [Accessed: January 2025].

20.5 The Institution of Structural Engineers. *Design for zero.* London: IStructE Ltd, 2021.

20.6 UK Green Building Council. *Embodied Ecological Impacts.* UKGBC, 2024. Available at: https://ukgbc.org/our-work/topics/embodied-ecological-impacts/ [Accessed: January 2025].

20.7 Expedition Engineering. *The Embodied Biodiversity Impacts of Construction Materials.* London: Expedition Engineering, 2024. Available at: https://expedition.uk.com/project/embodied-biodiversity-impacts-of-construction-materials/ [Accessed: January 2025].

20.8 Sealey-Huggins, L. 'The Climate Crisis is a Racist Crisis: Structural Racism, Inequality and Climate Change'. *The Fire Now: Anti-Racist Scholarship in Times of Explicit Racial Violence*. London: Zed Books, pp99–113. 2018. Available at: http://www.worldcat.org/oclc/1061285873 [Accessed: January 2025].
20.9 Bather, M. 'Viewpoint: business as usual?'. *The Structural Engineer*, 101(2), February 2023, pp40–41. Available at: https://doi.org/10.56330/TUQZ6963 [Accessed: January 2025].
20.10 Institution of Structural Engineers. *Guidance on ethics*. Available at: https://www.istructe.org/resources/guidance/guidance-on-ethics/ [Accessed: January 2025].
20.11 Ichioka, S. and Pawlyn. M. *Flourish: Design Paradigms for Our Planetary Emergency*. Charmouth: Triarchy Press, 2021.
20.12 Regensis Group. *Regenerative Development and Design: A Framework for Evolving Sustainability*. Hoboken: John Wiley & Sons, 2016.
20.13 Wahl, D.C. *Designing Regenerative Cultures*. Charmouth: Triarchy Press, 2016.
20.14 BBC News. *Volkswagen: The Scandal Explained*. Available at: https://www.bbc.co.uk/news/business-34324772 [Accessed: January 2025].
20.15 Norman, J. *et al*. 'How can we create an engineering industry while building nothing?' *The Structural Engineer*, 98(7), July 2020, p1. Available at: https://doi.org/10.56330/PAYC4580 [Accessed: January 2025].
20.16 The Institution of Structural Engineers. 'Nothing is better than something.' *The Structural Engineer*, 98(6), June 2020. Available at: https://doi.org/10.56330/NYMP4183 [Accessed: January 2025].
20.17 The Institution of Structural Engineers. *Circular economy and reuse: guidance for designers*. London: IStructE Ltd, 2023.

Part 6
Appendices

APPENDIX A

Introduction to probabilistic models and reliability analysis

Raffaele De Risi and Flavia De Luca

In this appendix, the reader will find some aspects related to the theoretical background needed to deal with data scarcity and compute the structural reliability of single structural components and more complicated structural systems. This appendix makes this book self-sufficient as it combines elements from several fundamental books (Section A.4) and summarises the key aspects.

A.1 INTRODUCTION TO UNCERTAINTY

Many engineering calculations are based on deterministic analyses, while significant uncertainties can exist in many input variables. Such uncertainties can be accounted for by adopting appropriate probabilistic approaches. Probabilistic approaches can be classical, frequentistic or Bayesian[A.1]. Classical approaches consider that any event has a given set of possible outcomes, all equally or unequally likely to occur. Frequentistic approaches consider probability as the relative frequency that can be observed in practice in many trials (e.g., experiments). Bayesian approaches consider probability as a quantification of the reasonable degree of belief. In this book, the term 'probabilistic' refers to a combination of these definitions.

Input data required for the design are often scarce and initially unknown; this is a strong argument for many to stay with traditional non-probabilistic approaches. However, data scarcity remains for non-probabilistic and probabilistic methods, therefore, uncertainties cannot be removed, and design methods remain, in essence, probabilistic. This is at the base of performance-based engineering (PBE). PBE is the design, evaluation and construction of engineered facilities that meet, as economically as possible, the uncertain future demands of owner-users and nature. The PBE framework is advantageous when the designer needs to deal with non-conventional structural systems, materials and loading.

Therefore, probabilistic methods are needed for fundamental engineering tasks such as modelling uncertain parameters, evaluating the results of an experiment or assessing measurement error. *Uncertainties* are design facets that are not perfectly defined as they are characterised by an imperfect, limited or unknown information level. They are conventionally divided into *aleatory* (i.e., intrinsic randomness), *epistemic* (i.e., lack of knowledge) and *model uncertainties* (i.e., imperfections and idealisations made in physical model formulations). All these uncertainties must be considered in any engineering decision analysis. From a practical point of view, uncertainties are usually summarised with uncertain parameters, also known as 'random variables'. Examples of random variables are the furniture's load on a floor, the material's compressive strength or the intensity/direction of a wind gust. Since any engineering system can be studied using empirical relationships and numerical/physical models depending on the uncertain input parameters, the system *performance* will also be a random variable.

Uncertainty modelling starts with investigating the available data[A.2], calculating the descriptive statistics (e.g., the sample mean, the sample variance and the coefficient of variation) and using graphical representations (e.g., histograms, cumulative frequency) (Figure A.1). Descriptive statistics alone are not enough, but they are an excellent tool for gaining a preliminary idea and communicating data uncertainties.

Figure A.1: (a) Histogram and (b) cumulative frequency plot

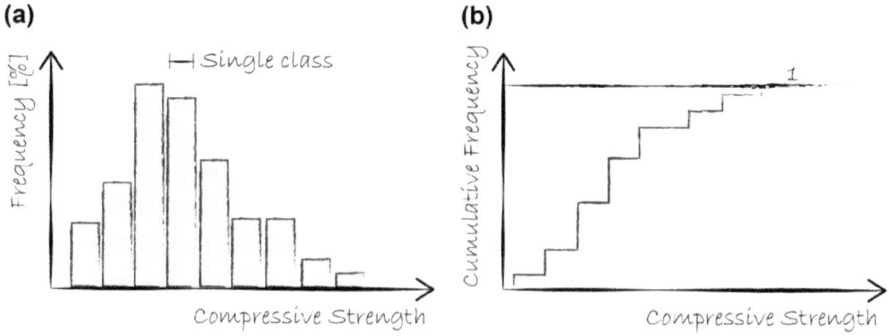

If the involved random variables can assume any real value, they are classified as 'continuous random variables', otherwise, they are called 'discrete random variables'. The probability that a generic continuous random variable X (usually represented with a capital letter) is lower than a specific value x (usually presented as a lower-case letter) is provided by the cumulative distribution function (CDF) (Figure A.2a):

$$F_X(x) = P(X \leq x) \quad \text{A.1}$$

where the capital letter F represents the CDF, the capital letter subscript (i.e., X) and the specific value in the brackets (i.e., x) are the generic random variable and the specific value, respectively. P is the abbreviation for probability; the term following P in the bracket is a statement. For the case of a CDF, such a statement is the inequality $X \leq x$.

The probability density function (PDF) (Figure A.2b) is the first derivative of the CDF with respect to the random variable:

$$f_X(x) = \frac{dF_X(x)}{dx} \qquad \text{A.2}$$

where the lower-case letter f represents the PDF, the capital letter subscript (i.e., X) and the specific value in the brackets (i.e., x) are the generic random variable and the specific value, respectively. Adopting a functional form for describing the uncertainties in the data allows simple handling and communication since only a few parameters need to be computed to completely define the PDF or the CDF. Such parameters are also known as 'moments of the distribution'.

Figure A.2: (a) Cumulative distribution function (CDF) and (b) probability density function (PDF)

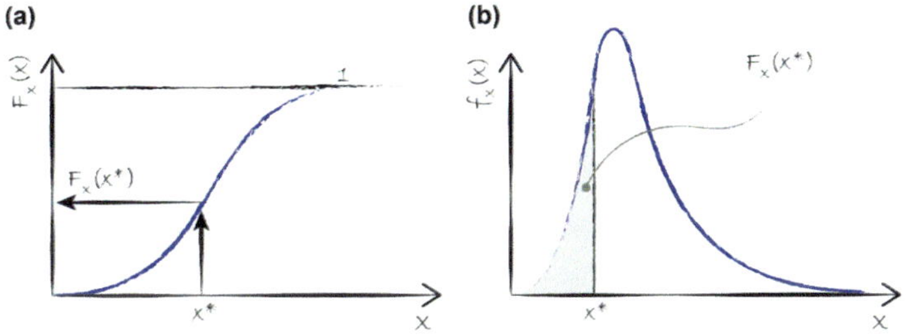

Several models (i.e., functional forms) exist (e.g., normal, lognormal and extreme value distributions), and their parameters (inferential statistics) can be computed with well-known methods (e.g., method of moments, maximum likelihood, probability plot). Chapter 10 presents some relevant probabilistic models.

The random variables discussed above are all univariate (or mono-dimensional), i.e., they refer to a single possible variable. However, engineering problems involve many variables. These variables are often correlated; examples are the compression strength and the elastic modulus of concrete or the wind velocity and snow loads on roof structures. To understand if any correlation exists between random variables, it is possible to create simple plots representing one variable against the other (Figure A.3a). Several correlation parameters can be calculated (e.g., Spearman's Rho and Pearson's coefficient correlation). If a correlation exists, correlation models should be defined, and the probabilistic model must be multivariate, i.e., consider more than one variable at once in joint probability density functions (Figure A.3b). Other examples of correlation are spatial and temporal correlations. Variables can be correlated in space because components close in space may have similar characteristics (e.g., two points on a concrete deck may have similar, if not coincident, compressive strengths). Similarly, variables can be correlated in time because chronologically close events may have similar characteristics as opposed to events further apart (e.g., wind gust time series).

Figure A.3: (a) Correlated random variables and (b) multivariate joint PDF

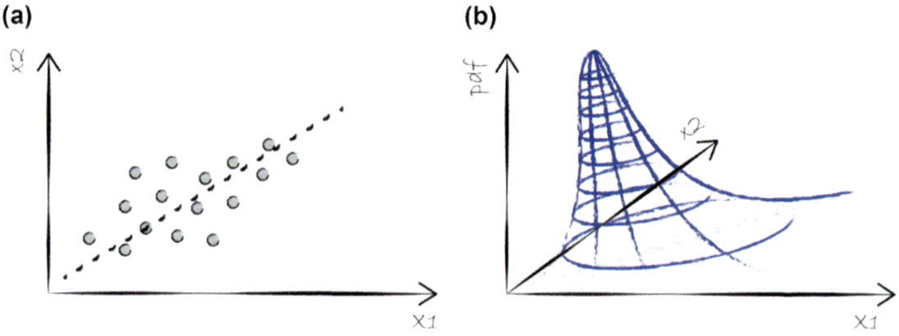

A.2 INTRODUCTION TO RELIABILITY

Once the uncertainties are characterised, a reliability analysis can be performed. The reliability analysis allows for incorporating and propagating explicitly uncertainties in the design, supporting the decision-making process. Any structural reliability analysis relies on availability of analytical/numerical/physical models of an engineering system that, excited by an external or internal perturbance (e.g., load, settlement), can return one or more engineering demand parameters (EDP), i.e., the demand or load effect D, that can be compared with one or more damage measures (DM), i.e., the capacity C. Load effect (demand) and capacity should be defined consistently, e.g., the shear force in a section and the corresponding shear strength. In reliability, failure occurs when the load effect (D) is larger than the capacity. Minor or major consequences can result when a component or a system fails; consequences are usually described using consequence classes or risk categories.

In the Eurocodes 'demand' is referred to as 'load effect'. As most readers of this book are familiar with these codes we have used this as the primary term. However, in reliability the term 'demand' is more generally used, so we have kept the D and have provided, from time to time, a reminder that 'load effect' is 'demand'.

To perform a structural reliability analysis, it is necessary to define a prescribed set of limit states, describing a condition of a system (or component) beyond which it no longer fulfils the relevant design criteria. This condition, as explained previously, is referred to as 'failure'. When the limit state is exceeded, the damage is described via the damage state, which provides an easy way to categorise the damage. The limit state separates the desired state from the adverse state of the system (or component). From an analytical point of view, the limit state is identified via a limit state function $g(\mathbf{X})$. The function $g(\mathbf{X})$ is the combination of load effect (D) and capacity (C). It depends on $\mathbf{X}$, where $\mathbf{X}$ is the vector of deterministic and uncertain scalar parameters in the problem (Figure A.4a). The failure is always defined as $g(\mathbf{X}) \leq 0$. The limit state function can assume two forms: safety margin (M) and safety factor (S) formulation, respectively[A.3–A.8]. The safety margin formulation is:

$$g(\mathbf{X}): \quad M = C - D \tag{A.3}$$

The safety factor formulation is:

$$g(\mathbf{X}): \quad S = \ln(C/D) \tag{A.4}$$

where M and S are random variable functions of C and D. In practical applications, the safety factor is C/D (and not the natural logarithm of the ratio), and failure occurs if C/D is lower than 1 (which is consistent with the definition of the logarithmic version of S). Chapter 6 provides a practical implementation.

As previously explained, both load effect (D) and capacity (C) encompass several sources of uncertainty, therefore the assessment of the performance with respect to the limit state function is probabilistic. Specifically, the probability of failure (P_F) is the multi-dimensional integral of the joint distribution of the random variables $f_x(\boldsymbol{x})$ over the failure domain (i.e., $g(\mathbf{X}) \leq 0$):

$$P_F = P\left(g\left(\mathbf{X}\right) \leq 0\right) = \int_{g(\mathbf{X}) \leq 0} f_{\mathbf{X}}\left(\boldsymbol{x}\right) d\boldsymbol{x} \tag{A.5}$$

The reliability of a system is the complement to 1 of the probability of failure P_F. This integral is non-trivial and can have a complicated integration domain. A simplified graphical representation of the probability of failure is shown in Figure A.4b, where the integral is the volume under the joint distribution based in the unsafe region. Classical integration schemes may not be suitable for solving these complicated multi-dimensional problems, and simulation-based approaches may be preferred in practical applications (Section 6.3).

Figure A.4: (a) Limit state function and (b) joint distribution of random variables

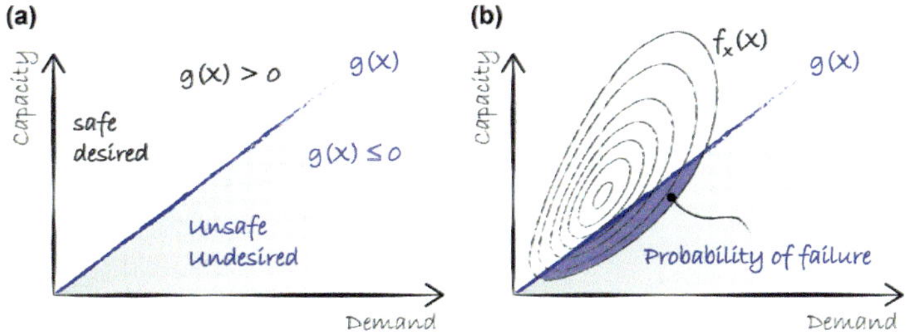

The relationship between the probability of failure P_F and the reliability (or safety) index β associated with a given limit state function is discussed in Section 6.4.

A.3 METHODS FOR CALCULATING PROBABILITY OF FAILURE

In Section 6.3, we use one method for calculating the probability of failure — namely the Monte Carlo method, which is a simulation-based approach. However, other methods can also be utilised. The following descriptions include examples based on the solution to the problem posed in Section 6.3.

A.3.1 Numerical integration

The first step for numerical integration is the computation of the joint distribution of the random variables, indicated with $f_x(\boldsymbol{x})$ in Equation A.5. In the example in Section 6.3, the two random variables, i.e., the force and the strength, are independent. Therefore, the joint distribution can be easily computed as the product of the two PDFs of the random variables ($f_x(\boldsymbol{x}) = f_{Fy} \cdot f_F$) indicated in Figure A.5 by the magenta and orange lines, respectively. The joint PDF is the 3D surface shown in Figure A.5; the contour plot is the planar representation of the surface.

The second step consists of the identification of the limit state function; for this example, the limit state function can be written with its safety margin (M) formulation as follows:

$$g: \quad M = W_{pl} \cdot f_y - F \cdot l \qquad \text{A.6}$$

where g is represented as a straight red line on the F-f_y horizontal plane in Figure A.5. The limit state function separates the safe from the unsafe region of the random variables.

Figure A.5: Single and joint PDF functions; *g* is the limit state function

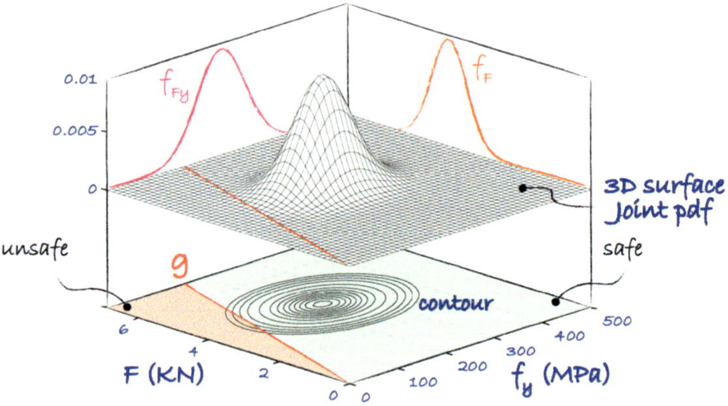

The third step consists of discretising the random variables in a finite number of intervals having a specific constant width df_y and dF; such discretisation will allow converting the integral of Equation A.5 into a summation. The last step consists of summing all the values of the joint PDF multiplied by df_y and dF only in the unsafe region of the random variables. For this specific example, the probability of failure is $2.9 \cdot 10^{-3}$, and the reliability index β is 2.76 according to Equation 6.10.

In general cases, the joint distribution of the random variables may be too complicated to be assessed. Therefore, the numerical integration may not be viable for cases involving many random variables, requiring different approaches and techniques to avoid simple mistakes and computational issues.

A.3.2 Approximate analytical methods

This family of methods approximates the problem as, in many practical cases, the limit state function is non-linear, and therefore, a linearisation using a Taylor series expansion is needed[A.9]. The most famous methods are first order second moment (FOSM), first order reliability method (FORM) and second order reliability method (SORM). Also, these methods rely only on a few moments, not the entire distribution. Finally, they are generally valid for random variables that are normal or lognormal.

This section will use the FOSM to solve the case study in Section 6.3. This method uses only the first two moments of the distributions of the random variables, i.e., mean and standard deviation, and requires a linear limit state function. Specifically, the probability of failure $P_F = P(g \leq 0)$ can be calculated by computing the moments of the limit state function presented in Equation A.6 using the property of the normal random variables presented with Equations 10.2 and 10.3 from Chapter 10:

$$\mu_M = W_{pl} \cdot \mu_{fy} - \mu_F \cdot l \qquad \text{A.7}$$

$$\sigma_M = \sqrt{W_{pl}^{\ 2} \cdot \sigma_{f_y}^{\ 2} + \sigma_F^{\ 2} \cdot l^2} \qquad \text{A.8}$$

The capacity, demand and safety margin (i.e., M) distributions are represented in Figure A.6a. The probability of failure can now be calculated as:

$$P_F = P(M \leq 0) = \Phi\left(\frac{0 - \mu_M}{\sigma_M}\right) = \Phi(-\beta) \qquad \text{A.9}$$

where β is the reliability index presented in its general form in Equation 6.10 and here shown in analytical terms for the safety margin formulation when all the variables are normal. Cornell[A.10] initially proposed this reliability index and defined it as the ratio between the mean value and standard deviation of the safety margin variable. From a graphical point of view, the probability of failure is the area under the probability density function of the safety margin distribution for negative values of the safety margin and β is the distance of μ_M from 0 in terms of σ_M (Figure A.6b).

Figure A.6: (a) PDF of the capacity, demand and safety margin variables; (b) zoom on the PDF of the safety margin, P_F and β

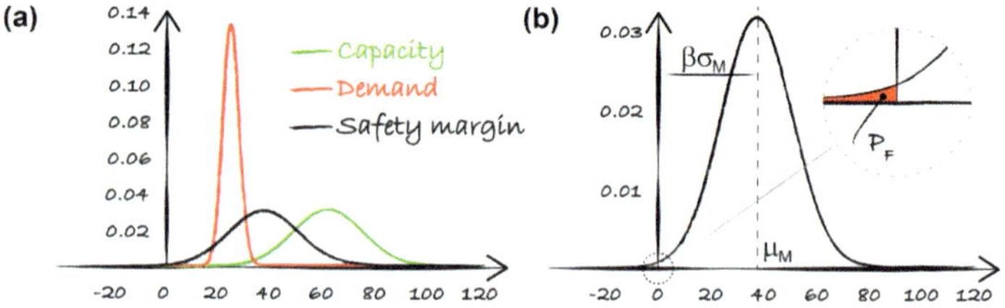

For this specific example, the reliability index β is 2.77, and the probability of failure is $2.8{\cdot}10^{-3}$. It is possible to note that the values are very similar to the numerical integration but are not the same (the probability of failure is given as $2.9{\cdot}10^{-3}$ in Section A.3.1).

If the random variables were lognormal, then the most convenient limit state function would be the safety factor formulation (Equation A.4). In such a case, Equation A.9 still stands, but instead of mean and standard deviation, there should be the log of the median of the safety factor and the logarithmic standard deviation of the safety factor. This is because the logarithm of the ratio is equivalent to the difference of the logarithms.

It is very important to note that when the FOSM is adopted to solve a reliability problem, the reliability index β may not be the same for different but mechanically equivalent formulations of the limit state function (e.g., $M = f_y - F \cdot 1/W_{pl}$). This is typically the case when the limit state function is non-linear. This issue is known as the lack of invariance of the reliability index. This problem was solved by Hasofer and Lind[A.11], by normalising the involved random variables (Figure A.7). Figure A.7a shows the joint PDF of the random variables and the contour plot already presented in Figure A.5.

Figure A.7: (a) Natural and (b) standard normal space

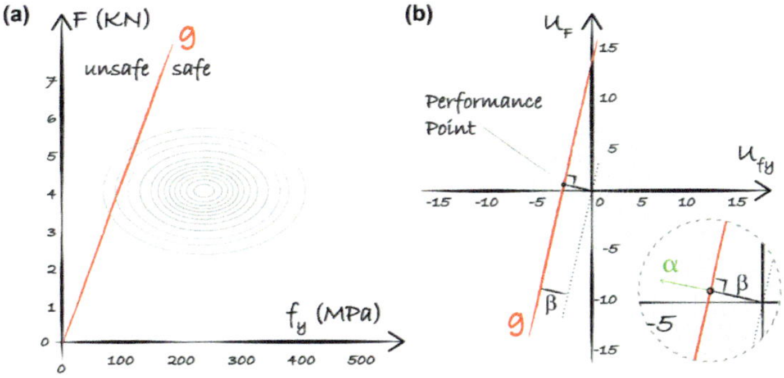

When the random variables are transformed into the standard normal space using Equation 10.4, any joint distribution of the variable becomes spherical around the centre of the axes (Figure A.7b). The transformation of variables also leads to transforming the limit state function. The shortest distance between the origin of the axes and the transformed limit state function is the invariant reliability index β (Figure A.7b). This number will not change with the form of the limit state function. The closest point to the origin of the axes on the limit state function is known as the performance point (or design point); such a point is representative of the random variables corresponding to the current probability of failure. Therefore, the performance point can be used to identify specific values of the random variables used for the design. The components of the versor

$\boldsymbol{\alpha}$ (Figure A.7b) normal to the limit state function at the performance point are known as sensitivity factors as they contain the sensitivity of β to uncertain random variables. For example, in the case study being considered, from the shape of the versor $\boldsymbol{\alpha}$ in Figure A.7b it is possible to conclude that the reliability is very sensitive to the yield strength and less sensitive to force.

A.4 FURTHER READING

This book has drawn on references representing a selection of valuable and seminal sources that have shaped the landscape of probabilistic models and reliability analysis in recent decades. A list of additional sources is provided here to complement the topics touched on in this book, and is categorised into four groups:

- Seminal papers covering the early days of reliability theory, which became milestones for design codes
- Papers and books covering advanced simulation techniques and surrogate models
- Other reliability texts not cited explicitly in this book complementing the overall discussion on specific topics
- A selected perspective on the classification of uncertainties in structural problems

A.4.1 Seminal papers

- Ang, A.H. and Cornell, C.A. 'Reliability Bases of Structural Safety and Design'. *Journal of the Structural Division*, 100(9), September 1974, pp1755–1769. Available at: https://doi.org/10.1061/JSDEAG.0003870 [Accessed: January 2025].
- Ellingwood, B.R. and Galambos, T.V. 'Probability-based Criteria for Structural Design'. *Structural Safety*, 1(1), 1982–1983, pp15–26. Available at: https://doi.org/10.1016/0167-4730(82)90012-1 [Accessed: January 2025].

- Ellingwood, B.R. *et al.* 'Probability Based Load Criteria: Load Factors and Load Combinations'. *Journal of the Structural Division*, 108(5), May 1982, pp978–997. Available at: https://doi.org/10.1061/JSDEAG.0005959 [Accessed: January 2025].
- Ellingwood, B.R. 'Probability-based Codified Design: Past Accomplishments and Future Challenges'. *Structural Safety*, 13(3), March 1994, pp159–176. Available at: https://doi.org/10.1016/0167-4730(94)90024-8 [Accessed: January 2025].
- Ellingwood, B.R. 'LRFD: Implementing Structural Reliability in Professional Practice'. *Engineering Structures*, 22(2), February 2000, pp106–115. Available at: https://doi.org/10.1016/S0141-0296(98)00099-6 [Accessed: January 2025].
- Ellingwood, B.R. 'Structural Reliability and Performance-based Engineering'. *Proceedings of the Institution of Civil Engineers — Structures and Buildings*, 161(4), January 2008, pp199–207. Available at: https://doi.org/10.1680/stbu.2008.161.4.199 [Accessed: January 2025].
- Davenport, A.G. 'Rationale for Determining Design Wind Velocities'. *Journal of the Structural Division*, 86(5), May 1960, pp39–68. Available at: https://doi.org/10.1061/JSDEAG.00005 [Accessed: January 2025].
- Galambos, T.V. *et al.* 'Probability Based Load Criteria: Assessment of Current Design Practice'. *Journal of the Structural Division*, 108(5), May 1982, pp959–977. Available at: https://doi.org/10.1061/JSDEAG.0005958 [Accessed: January 2025].
- Ghosn, M. and Moses, F. 'Reliability Calibration of Bridge Design Code'. *Journal of Structural Engineering*, 112(4), April 1986, pp745–763. Available at: https://doi.org/10.1061/(ASCE)0733-9445(1986)112:4(745) [Accessed: January 2025].
- Pugsley, A.G. 'Concepts of Safety in Structural Engineering'. *Journal of the Institution of Civil Engineers*,

36(5), March 1951, pp5–31. Available at: https://doi.org/10.1680/ijoti.1951.12755 [Accessed: January 2025].

- Ravindra, M.K. *et al.* 'Wind and Snow Load Factors for Use in LRFD'. *Journal of the Structural Division*, 104(9), September 1978, pp1443–1457. Available at: https://doi.org/10.1061/JSDEAG.0004987 [Accessed: January 2025].
- Shinozuka, M. 'Basic Analysis of Structural Safety'. *Journal of Structural Engineering*, 109(3), March 1983, pp721–740. Available at: https://doi.org/10.1061/(ASCE)0733-9445(1983)109:3(721) [Accessed: January 2025].

A.4.2 Advanced simulation techniques and surrogate models

- Au, S.K. and Beck, J.L. 'Estimation of Small Failure Probabilities in High Dimensions by Subset Simulation'. *Probabilistic Engineering Mechanics*, 16(4), October 2001, pp263–277. Available at: https://doi.org/10.1016/S0266-8920(01)00019-4 [Accessed: January 2025].
- Au, S.K. and Wang, Y. *Engineering Risk Assessment with Subset Simulation*. Singapore: John Wiley & Sons, 2014.
- Bucher, C.G. and Bourgund, U. 'A Fast and Efficient Response Surface Approach for Structural Reliability Problems'. *Structural Safety*, 7(1), January 1990, pp57–66. Available at: https://doi.org/10.1016/0167-4730(90)90012-E [Accessed: January 2025].
- Faravelli, L. 'Response-surface Approach for Reliability Analysis'. *Journal of Engineering Mechanics*, 115(12), December 1989, pp2763–2781. Available at: https://doi.org/10.1061/(ASCE)0733-9399(1989)115:12(2763) [Accessed: January 2025].

A.4.3 Other books

- Ayyub, B.M. and McCuen, R.H. *Probability, Statistics, and Reliability for Engineers and Scientists*. Boca Raton: CRC Press, 2011.

- Madsen, H.O. *et al. Methods of Structural Safety*. New York: Dover Publications, 2006.
- Thoft-Cristensen, P. and Baker, M.J. *Structural Reliability Theory and its Applications*. Berlin: Springer, 2012.

A.4.4 Uncertainty classification

- Der Kiureghian, A. and Ditlevsen, O. 'Aleatory or Epistemic? Does it Matter?'. *Structural Safety*, 31(2), March 2009, pp105–112. Available at: https://doi.org/10.1016/j.strusafe.2008.06.020 [Accessed: January 2025].

REFERENCES

A.1 Faber, M.H. 'Statistics and Probability Theory'. *Topics in Safety, Risk, Reliability and Quality*, 18, 2012.

A.2 Benjamin, J.R. and Cornell, C.A. *Probability, Statistics, and Decisions for Civil Engineers*. New York: McGraw-Hill, 1970.

A.3 Ditlevsen, O. and Madsen, H.O. *Structural Reliability Methods*. Chichester: Wiley, 1996.

A.4 Pinto, P.E *et al.* 'Seismic Reliability Analysis of Structures.' *Journal of Earthquake Engineering*, 9(6), 2004, p943. Available at: https://doi.org/10.1080/13632460509350573 [Accessed: January 2025].

A.5 Ang A.H-S. and Tang, W.H. *Probability Concepts in Engineering*. Hoboken: John Wiley & Sons, 2007.

A.6 Nowak, A.S and Collins, K.R. *Reliability of Structures*. London: CRC Press, 2013. Available at: https://doi.org/10.1201/b12913 [Accessed: January 2025].

A.7 Melchers, R.E. and Beck, A.T. *Structural Reliability Analysis and Prediction*. Hoboken: John Wiley & Sons, 2017.

A.8 Der Kiureghian, A. *Structural and System Reliability*. Cambridge: Cambridge University Press, 2022.

A.9 Chamberlain, A. *An Easy Way to Remember the Taylor Series Expansion*. Medium, 2016.

A.10 Cornell, C.A. 'A Probability-based Structural Code'. *American Concrete Institute Journal Proceedings*, 66(12), January 1969, pp974–985. Available at: https://doi.org/10.14359/7446 [Accessed: January 2025].

A.11 Hasofer A.M. and Lind, N.C. 'Exact and Invariant Second-moment Code Format. *Journal of the Engineering Mechanics Division*, 100(1), February 1974, pp111–121. Available at: https://doi.org/10.1061/JMCEA3.0001848 https://doi.org/10.1061/JMCEA3.0001848 [Accessed: January 2025].

APPENDIX B

Reliability: a systemic non-linear time-dependent problem

Raffaele De Risi

In almost all practical cases, the limit state function is non-linear. This means that to apply some of the analytical methods presented in Chapter 6, the limit state must be differentiable (i.e., derivatives can be computed) so that approximation (e.g., Taylor expansions[B.1]) can be performed. However, the simulation-based procedure can deal with the non-linearity easily without approximations and without having a differentiable limit state function. Also, the limit state function can change with time; therefore, the reliability problem, in general, is not time-invariant. This is due to the stochastic nature of the stressors and the ageing properties of construction materials. Figure B.1a shows the schematic time-dependence of load effect (demand) and capacity for a generic reliability problem. It is possible to observe that the capacity tends to decrease, and the load effect tends to increase. Also, the probability density functions of the load effect and capacity become wider and flatter. The time-dependent reliability problem leads to an increasing probability of failure with time (Figure B.1b).

Figure B.1: (a) Time-dependent reliability problem and (b) time-dependent probability of failure

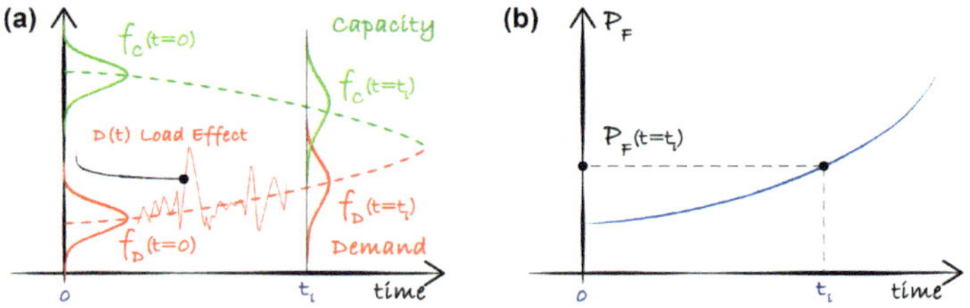

The failure event constitutes the outcrossing of the time-variant limit state surface. Such outcrossing is also known as the 'first-passage' (or first-excursion) problem. Assuming the probability of failure independent of time, the probability of observing the failure in any time interval [0, t] can be computed as:

$$P(T<t)=F_T(t)=1-\prod_{i=1}^{t}\left[1-P_F(t_i)\right] \qquad \text{B.1}$$

The CDF presented in Equation B.1 can be used to compute a conventional measure used in classical time-dependent reliability theory, i.e., the hazard function h(t). The hazard function provides the probability that a failure can occur in the time interval [t, t + dt], knowing that the failure did not happen before t. For this reason, the hazard function is also known as 'conditional failure rate' and is equal to:

$$h(t)=\frac{f_T(t)}{1-F_T(t)} \qquad \text{B.2}$$

where $f_T(t)$ is the PDF computed from the CDF $F_T(t)$. The hazard function for a realistic structure looks like the one shown in Figure B.2. It has a classical bathtub shape that can be divided into three parts:

- A burn-in period of early failure due to construction/human/design errors
- A stable period in which the structure is under service and the failure is due mainly to chance
- A final burn-out period where the increasing failure rate is due to the system degrading with time

Figure B.2: A prototype 'bathtub' hazard function

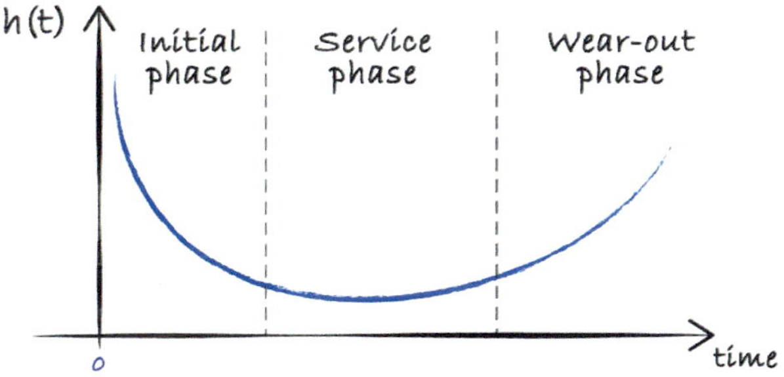

A note on the reliability index

Equation B.1 can be used to convert the reliability index obtained for one year to any reference period. Specifically, assuming that the failure is independent year-on-year, the reliability index for a specific reference period is:

$$\beta(t) = \Phi^{-1}\left[\Phi(\beta_{1\,year})^{t}\right] \qquad \text{B.3}$$

This equation can be used to understand the reliability index values presented in Section 6.4. However, Equation B.3 is valid only for independent failure events in subsequent years. If we assume that t_0 is the period for which the failure events are independent, Equation B.3 can be reformulated as:

$$\beta(t, t_0) = \Phi^{-1}\left[\Phi(\beta_{1\,year})^{t/t_0}\right] \qquad \text{B.4}$$

where $t/t_0 \geq 1$ is the number of events potentially leading to failure. The time t_0 can be assumed equal to one year when we are concerned about human life and for reliability problems governed by actions such as traffic or environmental effects. A time t_0 of five to ten years can be considered when the reliability problem is dominated by a sustained load, such as that in office buildings. If the reliability problem is not affected by time-variant issues (e.g., the case of reliability problems dominated by permanent loads)[B.2] $t_0 = t$. It is possible to conclude that the hypothesis of independent failure events needs to be appropriately checked before selecting a specific reliability level.

The systemic problem

The discussion presented so far refers to a simple analysis of single components. However, structural systems are made of interconnected components and members. Therefore, the failure of a single component may or may not necessarily correspond to the failure of the entire system. This is because structures are redundant. To approach such a complicated study it is generally appropriate to introduce two idealised systems[B.3]: series and parallel systems (Figure B.3).

Figure B.3: Reliability problems for two random variables x1 and x2: (a) series system and (b) parallel system

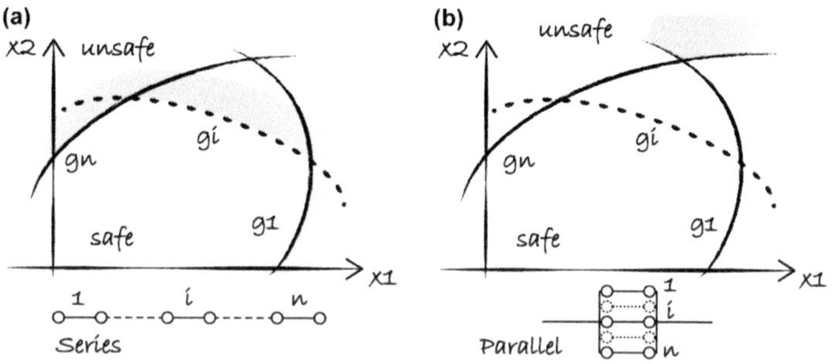

For a series system, the failure of a single component leads to failure of the entire system; therefore, the system's capacity depends on the minimum capacity of its components. For this reason, the series system is also known

as the weakest link system. For a parallel system, all the components must fail for the global failure of the system to occur. For a series system, the unsafe domain is the union of the unsafe domains of each component (Figure B.3a); conversely, for a parallel system, the unsafe domain is the intersection of the unsafe domains of all the components (Figure B.3b).

Suppose it is possible to assume that the *n* capacities of each single component of the system are statistically independent. In that case, the systemic probability of failure can be calculated for a series and a parallel system using Equation B.5 and B.6, respectively:

$$P_F = 1 - \prod_{i=1}^{n} \left(1 - P_{F,i}\right) \quad \text{B.5}$$

$$P_F = \prod_{i=1}^{n} P_{F,i} \quad \text{B.6}$$

where n is the number of components, and $P_{F,i}$ is the failure probability of each single component. For a series system, the probability of failure is always larger than the probability of failure of the single components. Conversely, for a parallel system, the probability of failure is always smaller than the probability of failure of each single component.

For complicated structural systems, it is very important to distinguish between brittle and ductile failure mechanisms. The nature of the failure mechanisms of each component may lead to a redistribution of the demand in the system; such a redistribution should be taken into account. Also, the formulations presented here are valid if the components' capacities are statistically independent. If any form of correlation exists, this should be properly accounted for[B.4].

Finally, a real structural system can be idealised with a combination of series and parallel systems. These general systems are known as 'cut-sets' and 'tie-sets' according to the nature of the interaction between the components, and according to the nature of the failure mechanisms[B.5].

REFERENCES

B.1 Chamberlain, A. *An Easy Way to Remember the Taylor Series Expansion*. Medium, 2016.

B.2 Sykora, M. *et al.* 'Target Reliability for Existing Structures Considering Economic and Societal Aspects'. *Structure and Infrastructure Engineering*, 13(1), August 2016, pp181–194. Available at: https://doi.org/10.1080/15732479.2016.1198394 [Accessed: January 2025].

B.3 Der Kiureghian, A. *Structural and System Reliability*. Cambridge: Cambridge University Press, 2022.

B.4 Nowak, A.S and Collins, K.R. *Reliability of Structures*. London: CRC Press, 2013. Available at: https://doi.org/10.1201/b12913 [Accessed: January 2025].

B.5 Ditlevsen, O. and Madsen, H.O. *Structural Reliability Methods*. Chichester: Wiley, 1996.

APPENDIX C

A short guide to testing

James Norman and Scott Boote

While this book is aimed at challenging engineers to think differently about uncertainty and risk and is not intended as a 'how to', it seems appropriate to introduce some of the practical testing elements, so that engineers can begin exploring this important part of understanding uncertainty. The following is not an exhaustive guide for designing, specifying, overseeing and, in some cases, running all tests but, instead, a set of suggestions that we use with undergraduates when they start physical testing research projects. It is based on our own experience designing and carrying out tests and knowledge gained when supervising undergraduate and PhD projects which involve physical testing.

C.1 PLANNING

Testing, like all engineering activities, requires careful design, planning, costing and risk assessment. It also requires excellent communication to clients about what you are doing and what can go wrong.

While it is impossible to give detailed instructions on how to approach a test, the following sections offer some guidelines of things to consider before, during and after the test(s).

C.2 BEFORE TESTING

Before you begin detailed planning, we suggest you consider the different facets of testing outlined in Chapter 11.

Understanding what it is you need to know is key. Spend time thinking about what effect different behaviours may have on the performance of your structure. This may not be immediately obvious and some form(s) of analysis may be required. It is easy to assume that failure or ultimate strength would be the most important parameter, but is that really the case? A sensitivity analysis can help. For example, if designing a doubly-curved compression-only structure, the strength may not govern the design. Instead, the failure may be due to buckling. As such, we may be more interested in the stiffness and dimensional stability of the material. Significant changes in shape will lead to changes in thrust lines, which may, in turn, lead to geometric instability, buckling and failure.

Once we are sure testing is necessary and we know what we need to test for, it is always good practice to try things out and carry out an initial study. Even if this simply means making or acquiring a sample of the material you are interested in — that initial visual and tactile investigation will be of value. We would go further and suggest even if you don't need to test a material, playing with it and exploring it is still a valuable exercise for any engineer.

C.2.1 Risk assessment

Of course, you should carry out a risk assessment before you do anything, even playing with a material. In fact, with testing you should risk assess continuously.

To an extent, the reason for carrying out testing is to find out what you don't already know. This level of unpredictability is the very reason you are carrying out testing, but this could also give rise to unpredictable hazards. A risk assessment should always be seen as a live document, but with testing even more so. You should update it before and after every test, and if necessary, during the test, too.

You should carefully consider who is involved. You may invite non-technical people to the test for whom the risks may not be as obvious as they are to construction professionals. To aid with this, test houses and universities will have their own checklist of risks to consider alongside your review process.

C.2.2 Design the experiment

Next, you need to design the experiment.

What is the physical set-up? What equipment do you need? What is the specification? Are there space or capacity requirements? Do you need anything fabricated beforehand? Who will be responsible for doing this? Will it need to be designed and drawn?

The test must isolate the specific behaviour that needs to be investigated. Consider all the effects of your test set-up, how this could affect the results and how the set-up interacts with the specimen. For example, if carrying out a shake table test, will the natural frequency of the support frame interact/interfere with your results? If a tension test is required, how much does the testing equipment elongate and will this affect your readings? If a compression test is to be carried out, will the specimen or support fixings buckle (Figure C.1)?

Figure C.1: Bamboo tested under cyclic load — the top plate, which is only required to support the specimen, has buckled under axial compression

C.2.3 Design the testing regime

Once you have designed the physical set-up, you need to design the tests.

How many repetitions you will undertake will depend on several factors. If you have some prior knowledge, the minimum number is one, but if not, it is four. Remember, as the number increases, the uncertainty reduces, but the cost and time taken may increase, so a balance must be struck.

If we are looking to do a comparative study or baseline testing — for example, looking at the impact of the addition of ground granulated blastfurnace slag (GGBS) on the colour of concrete, as we did for the John Henry Brookes building at Oxford Brookes University, alongside the different test specimens — it is important to have a control sample.

In addition to the number of samples, you need to consider how much load you will apply and how quickly. Do you want to step the loading (have some pauses) to see if there are any creep effects? Do you want to cycle your loading? This can help if the behaviour of the specimen cannot definitively be said to be purely elastic. Do you need to apply some pre-loading to allow for bedding-in of the specimen before you carry out the full test? This can be useful to avoid spurious readings at the start of a test. You should produce a table and graph with the load or displacement profile beforehand. As with the number of tests, this can have a significant impact on the time taken to complete the experiment (and often the cost, too).

If you are carrying out a different type of test, for example, measuring the abrasion resistance, how will the surface be abraded? How many cycles will be applied and what pressure will be used? Or if running a long-term test, how much load will be applied, and when will measurements be taken? If checking the post-fracture strength of structural glass, what will the testing regime be? All tests, whether in the lab, out on-site or in your office, should have a plan that considers all of the potential variables, and controls/adjusts them accordingly.

Before inventing this from scratch, check to see if there is a test standard you can apply. If not, are there similar tests published in journals that you can use?

C.2.3.1 Check the expected maximums and minimums

To ensure useful data can be obtained you will need to know the anticipated failure load (and other maximum/minimum values, such as deflection, frequency, shrinkage) before you design the test, which can be tricky as you won't know this until you've tested it. This gives you the option of either doing a rough test for yourself or using existing information to make a best guess.

You can then sensibly size your instrumentation, test rig, etc. Most instruments have a sweet spot where they give accurate outputs. As they approach the limits of their capability, results can become less reliable or you may simply run out of capacity/travel. At the other end of the spectrum, the result may be so small as to be highly susceptible to error, either electrical 'noise' if you are using sensors, or if you are taking manual readings, you may not be able to read results with sufficient accuracy (Figure C.2). You can only find this middle ground and use the correct sensors/measuring tools if you have an expectation of what will happen.

Figure C.2: Testing range of an LVDT and the impact of noise on the output

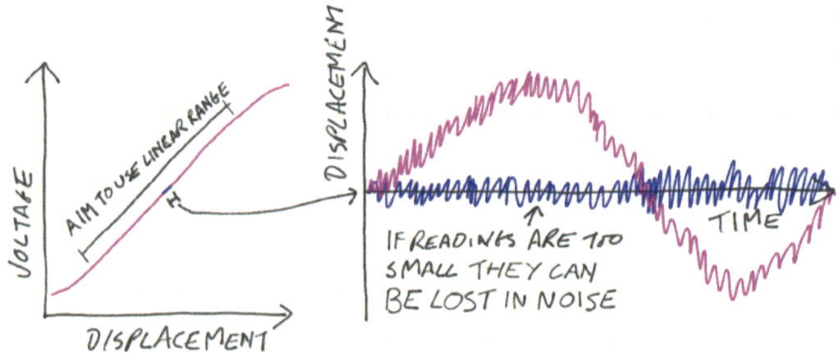

C.2.3.2 Plan what to measure

Consider what you are measuring and why. For every sensor or measurement, you create extra data. This is a good thing (the more data we have, the more we know) and a bad thing (the more instruments we use, the longer it takes to set up and the longer it takes to process the data).

There is nothing worse than finding out that while you have enough data to confirm how a material or system behaves, you can't quantify this due to a lack of data and, therefore, need to go back and do more tests. However, there is also a deep frustration in having 256 channels of data and sampling it at 1,000Hz, and it takes hours just to open the file!

Getting the right amount of data is a fine balance. Discussing the appropriate amount of instrumentation with an experienced technician may be very helpful.

C.2.3.3 Understand the accuracy of measurement

No instrument will give a perfect reading, as every instrument has an associated error. To understand the size of this error, we can take a known quantity, measure the same quantity with our instrument and see the difference. This process is known as 'calibration'.

The known quantity will itself have a degree of accuracy. Many of the most precise unit measurements (a kilogram, a metre, etc.) are stored at the National Physical Laboratories[C.1]

Many lab-based instruments will come with calibration certificates — these will say when it was last calibrated, when it next needs to be calibrated and the accuracy of the measurement.

For many instruments, however, especially the type you might use in your own workshop, you may need to calibrate the instrument yourself. This is actually a really good way to ensure you know what the output of the instrument is and how it works. For displacement-based readings, this could involve taking measurements with, say, a dial gauge and ruler and comparing the output. Remember that your ruler is probably no better calibrated than your dial gauge — but it will alert you to gross errors or your own misunderstanding. For an accelerometer, you can perform a 2g test. First, place it flat on a horizontal surface (ideally set with a spirit level) then turn it over. The difference in the reading should be 2g. For force, you can apply a known mass and check the readings. A commonsense approach should be applied, but calibration is helpful, even if it simply prevents you from using a faulty instrument in your experiment.

The calibration should ideally be carried out for the full range of readings you expect from the experiment, as the extreme values may be less linear than the central portion of the response. This is especially true for non-mechanical forms of measurement (Figure C.2).

For some instruments, such as strain gauges, calibration may be challenging as they may be single-use; once you have used them for calibration, they can no longer be used!

The most important consideration at this stage is whether the magnitude of the error will be a problem. If the error is the same magnitude as the readings you are hoping to take, then the instrument you plan to use is clearly not well-suited. Ideally the error should be significantly less than the reading, but defining this magnitude is hard as it also depends on the stage and purpose of testing and the cost of increasing the accuracy.

Understanding errors before you conduct the test is very important. Nothing is worse than carrying out an error analysis of your results after the experiments, to find that the data may be much less useful than hoped. Therefore, it is worth investing time early thinking about this, as it may prevent wasting time and effort later.

C.2.3.4 Familiarise yourself with the test space and the people who work there

Similar to working on an existing building and finding a caretaker who knows everything about the site for the last 50 years, finding a technician who has worked in the lab for years can be gold. They will know so much and will be full of helpful practical advice. Before testing, book a visit to the lab a few times. Go with the aim of observing and learning. Take notes, ask pertinent questions, learn from other people's mistakes and identify what is available and what is not.

In addition, be mindful of storage and testing conditions. What is the temperature and humidity level? Is it inside or outside? Make sure that your specimen isn't adversely affected by these conditions and that you record them and, if necessary, take readings.

C.2.3.5 Plan to capture all the information

Plan how to capture all the information and who is responsible for what. Take more people than necessary, ask someone to take photos and set up a video (a simple tripod and smartphone should be adequate). Assign another to take readings and a third just to observe. Of course, these don't all have to be people from your company, but generally, people enjoy watching experiments. If you are logging data by hand, make sure you take rigorous notes. Don't be tempted to skip any steps because a test appears to be following a pattern observed in previous tests (Section C.5.3). If you are using a data logger make sure you know what type of output you'll get and if you can access it/use it? What frequency are you recording at? If a frequency is too large you will end up with far too much data but if the frequency is too small you may miss a critical moment —

ideally you want a minimum of ten data points between each point of load application and 100–1,000 data points in total, although it really does depend on what you are measuring. If you are looking at anything dynamic, beware of 'aliasing' which is where a lack of data points leads to misinterpretation of data (Figure C.3).

Figure C.3: Phenomenon known as aliasing

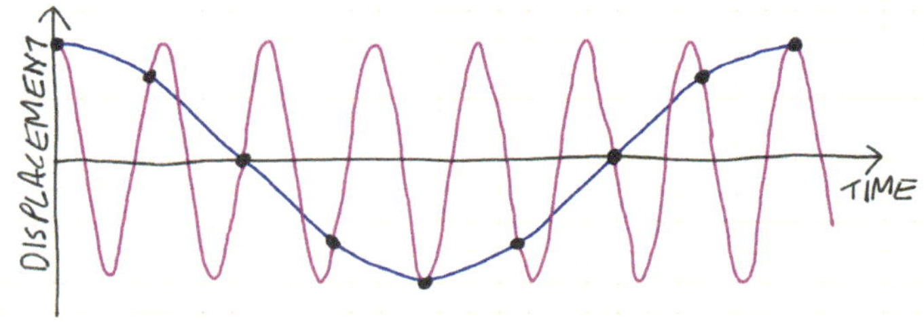

C.2.4 Cost the project appropriately

Once you have designed the test, you need to have it costed. To do this, it may be helpful to provide specifications. This should cover the following:

- Who is supplying what? (i.e. who is making the samples?)
- How many tests are needed, who is setting up/dismantling them and what information is required at this stage? It is much quicker/cheaper to crudely destroy an old specimen than to clear away failed material carefully
- The test set-up to include instruments, fixing plates, data logging, method of load application, etc.
- The test schedule — including pauses and moments for review of data
- The test data — including film, photos, notes, data, data checks, whether data needs to be visible during a test and how it will be recorded and shared afterwards
- The day rate for extra days if the tests take longer than planned (they probably will) and the impact of this
- What happens to the samples at the end of testing and any other unused samples?

- Anything else you want to include — who provides the hot chocolate, what safety equipment is required, what changing facilities are provided, etc.?

The testing can then be costed against the specification, and extras can then be charged as and when necessary. Although small changes will probably not be charged, extra day(s) probably will be.

Part of the costing of the project is also working out who will be paying for it. Most clients will be quite surprised to discover that testing is not a normal part of the fee, especially if you suggested the material and are now asking for extra fees to test it. It is vitally important to communicate early and continuously with the client. If they are proposing the material, or if the proposal comes from the architect, it may be easier to persuade them of the need to pay for testing.

You also need to communicate the risks — 'We may spend money on testing only to discover it doesn't work!' You will also need to agree on who owns the intellectual property and what happens with the data after the project. If the client is paying for the tests, it is not unreasonable for them to own the data (at least in part), so this should be agreed upfront. Ideally, the conversation will be open and continuous and will cover all these areas.

C.2.5 Some useful instruments and what they are used for

When testing, you will want to take some measurements. There is a wide range of ways to take these measurements. In Table C.1 we have listed some general instruments that might be appropriate. In addition, if you search for these online, a bit like finding a book in a library, you will find that the shops that supply them will offer several alternatives not mentioned due to the need for brevity. Investigate fully and don't assume that just because something isn't listed in Table C.1 it is not available.

Table C.1: Some useful measuring instruments

Instrument	Useage	Is it cheap or expensive? Where do you source them?	Manual or electronic	Things to note
LVDT (linear variable differential transformer or linear variable displacement transducer)	Measures displacement	Expensive Available at most academic test houses and commercial testing facilities	Electronic	Accurate but delicate and expensive
Linear motion potentiometer	Measures displacement	Cheap Electronics stores	Electronic	Less accurate than LVDT
Dial gauge	Displacement	Cheap Hardware store	Manual	Limited range — check range suitable for application
Callipers	Distance	Cheap Hardware store	Manual	Can be used to measure thickness as well as taking strain measurements with stick-on markers
Ruler	Distance	Cheap Hardware store	Manual	May change length with temperature
Stick-on markers for concrete	Measuring strain and changes to crack widths	Cheap Online retailers	Manual	These are one-use only and measurement is generally manual
Strain gauge	Strain	Cheap but disposable so cost can escalate if you use a lot Electronics stores	Electronic	Installation can be fiddly. If damaged will stop giving readings
Digital image correlation	Full surface strain field	Expensive May be available at test house	Electronic	Requires very careful set-up and specialist support
Tell tale	Crack monitoring gauge	Cheap Hardware store	Manual	Useful for creep/cracks where you want a quick visual check

Instrument	Useage	Is it cheap or expensive? Where do you source them?	Manual or electronic	Things to note
Load cell	Load	Expensive Available through testing house	Electronic	Often integrated into testing machines
Force washer (specific load cell)	Measuring force through a bolt	Expensive Available through testing house	Electronic	
Spring balance	Load	Cheap Hardware store	Manual	Easy to use, only measures tension, but extends in the measuring process so could influence experiment
Scales	Weight	Cheap Hardware store — different sizes available which offer different ranges and levels of accuracy	Manual — digital display but need to write down output	Take care in selecting correct scale for range of values
High output linear accelerometer	Acceleration	Expensive May be available at test house	Electronic	Very accurate but delicate. Will break if dropped
Accelerometer MEMS (micro electro mechanical system)	Acceleration	Cheap Electronics stores	Electronic	Less accurate but more robust
Drawstring transducer (also known as 'string pot')	Displacement	Expensive Available at most academic test houses and commercial testing facilities	Electronic	The string is attached to the specimen. This has two advantages, firstly the string is flexible so it is less likely to affect readings and secondly they are less liable to brake if a specimen fails, as the string is unable to buckle. Less accurate than LVDTs

In addition, a number of tools are available on-site, from slump cones and laser levels to theodolites, co-ordinate measuring machines (CMMs) and a wide array of other tools. These may be used for measuring in a lab or in your workshop.

As mentioned, Table C.1 is not all-inclusive — a quick search for any of the following will produce another list of instruments you could consider: video extensometer, clip gauge, inclinometers, thermocouple, IR cameras, rotary encoder, torque transducer, differential global positioning system (DGPS), laser monitoring. The main question is: what are you trying to measure? Once you have determined this there is almost certainly a device out there that is able to measure it.

C.3 ON THE DAY

Ensure you have read the risk assessment and all measures are in place.

Arrive early and don't rush.

Be prepared for long waits. If you are outside (or in a warehouse or any lab even) dress warmly, take a hot drink and have some food available. If you are cold or tired, you are more likely to rush things, which is less safe and also means you might miss something.

Check all the instruments — are they giving accurate readings? When were they last calibrated? Nothing is more frustrating than spending the whole day testing to discover that one of your instruments has a short circuit and the output is nonsensical.

Do a trial run. Ideally, this will be a few days before, so you can review the results and update the experiment, if necessary. If that is not possible, make sure to include a test to ensure all your working assumptions are correct. If not, have a plan B in place (if the specimen is stronger, what will you do, and if it is weaker, what will you do?).

Check the data — did it all record OK? Open the output file and see if you know what each column represents. If you use more than one of the same instruments, make sure you know which data belongs to which instrument. Make sure you know what negative data represents. If possible, carry out simple calibrations.

Review the risk assessment and review the plan.

Have a short break — making time for reflection will help you get the most from the day. Make notes as necessary.

Set up the first full test. Check it. Is the specimen positioned in the right way? Pause (set a countdown timer for five minutes). Double-check the specimen. Once the experiment has started, you can't stop, so don't rush.

Begin the experiment. Take lots of photos/video. Make lots of notes, recording even the most obvious details. Observe. Once the experiment has finished (depending on the failure) pause and don't immediately remove the specimen.

Review the plan and the risk assessment. Make any necessary updates.

Check you have all the photos you need. Do you want to look at anything on or inside the failed specimen (and is this covered by your risk assessment?). Make sure you gather everything you need, as once it is gone it is gone.

Now repeat. Sequentially number the experiments; this will help with the automation of your post-processing after the event. If you are taking photos/film, consider capturing the test number and purpose so you can easily associate them with a specific test later on.

Stick to the plan, don't rush, and check the data after every test. Have all the instruments worked?

Ideally build in some extra time at the end. Have you noticed anything that requires further investigation?

Once the day is finished make time, ideally as soon as possible after the event, to process both the data and the events/observations noted during testing.

C.4 AFTER THE EVENT

Once your tests are completed, you will need to analyse your data (Section C.5), but you should also take time to record and file as much information as possible. Start by writing up your notes. Try to recall everything you noticed; from any issues you had with creating samples and setting up the experiments to anything you noticed during the test. You should already have this in note form, but over time, these notes will be harder to interpret.

Also, file any photos and videos you may have. Again, try to link them to tests as precisely as possible. Use time stamps to record the order of events. This could be very helpful after the event as you try to make sense of the data.

Finally, try to make your data processing as traceable as possible. You may well start with a CSV file with thousands or millions of data points. Try labelling each file so you know which is which. Add headings. If you start to process raw data, explain what you did at each step so that, if necessary, you can repeat it.

Remember that all of this information is useful — possibly to you again in the future, perhaps to someone else in your company and maybe to another engineer. Ideally, you should consider sharing it so that others can learn from what you have done. The more we share, the more progress we can make in changing how we design!

C.5 ANALYSING THE RESULTS

Once testing is complete and all the data has been downloaded, you will want to go through the process of making sense of it. In many ways, this is the most complex part of the process as you need to try to make sense of the outcome without biasing the results. It is helpful, as you start this process, to have some idea of what you were trying to achieve, e.g., is the test data a single concrete cube result to confirm 28-day strength, or a suite of tests on a brand new material for which you have no preconceived idea how it will respond? Does the data raise serious concerns? Or does it match exactly what you were expecting?

In Sections C.5.1–C.5.4 we discuss what you must consider as you analyse the data. Two key questions to answer as you do this are:

- How does the data from the test fulfil the test requirement?
- Does the data from the test (and data can be anything from observations to numerical outputs) suggest we need to reconsider the original expected outcome?

C.5.1 Interpreting results

Next, you will need to make sense of the data. This can be more difficult than you first imagine. As we saw in Chapter 10, there are several different idealised responses of a structure, whether stress-strain curves or abrasion descriptions. Trying to match our own measurements to these can be challenging.

Consider the force–displacement diagrams for a bamboo connector in Figure C.4 — these are the outputs of the testing seen in Figure 11.4. On the surface, we may imagine them all to be the same, and yet we have six seemingly different outcomes that we need to interpret. On some of the tests, we see bedding-in — a region at the start of the test where there is displacement with very little load (C and, to a lesser extent, A). Is this due to lack of fit in the experiment, or is it an expected property of the material? We then see an 'elastic' response, but even here, this is complex. Some lines look straight and have an easily measurable stiffness (A, B, D); others are more curved (C, E, F). When does the elastic range stop? Is it really elastic (which means that when we remove the load it will return) or is it permanently deforming? Then there is the transition into what we assume is a 'plastic' range. Far from being well-defined, in all cases apart from D, the exact position of this change is hard to determine. However, deciding where this occurs will define the allowable stress for SLS cases and determine the measure of ductility. The one value that is clear is the maximum value. But is this the right value to

Figure C.4: Force–displacement diagrams for six different bamboo screw connections

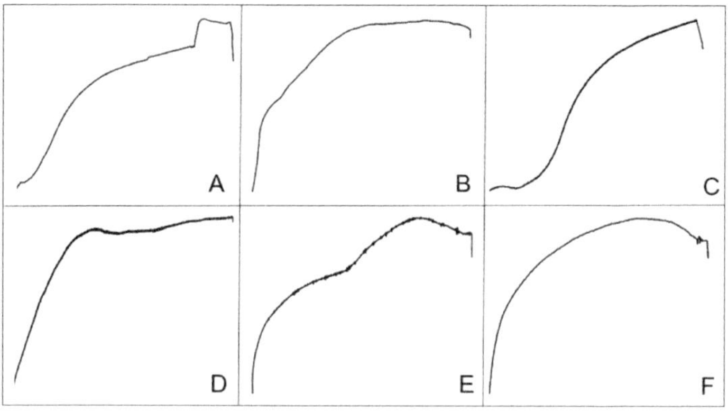

design for, especially if it occurs, as is the case of C, just before failure? And when does failure occur? Is it the point before the sharp reduction in load? In the case of C, the final fall-off of load is less vertical — does failure occur at the left or right of the final diagonal portion?

Learning to interpret this data is both a skill and an art form. It is a skill because the more you do it, the more confident you will become with the outcome. It is an art form because interpreting the data is part of a complex system with many dimensions. Understanding the art of interpreting the system will help you learn which details matter.

Unpicking the data in Figure C.4 reveals some very important things to consider. The plots don't reveal the whole story. To obtain a complete picture, it's necessary to refer to other data, such as photos and your notes, to see what is really happening. The plots represent not one failure type but multiple. They all occur at roughly the same load depending on screw diameter, head type, material thickness, connector plate type, etc. By understanding the failure type, we can more accurately appreciate what the data is telling us.

We should also be mindful of why we need this data in the first place. If we are worried about stiffness, not strength, we may want to spend more time thinking about the gradient of the line rather than the peak value. The bedding-in zone may be of particular concern if it is real and could lead to changes in geometry. For example, in a vaulted structure, it may persuade us that we need to preload our own structure to ensure the geometry is correct at the point of completion.

If testing standards were used, we might also turn to these for further advice. They may contain helpful advice not just on the design of an experiment, but the interpretation of the data. The one thing we should avoid is too quickly compressing the data to single failure stress or stiffness values, and not spending time with the complexity. For example, it would be easy to collapse the data into a distribution of maximum values from each of the graphs in Figure C.4 without really thinking through the differences in behaviour.

C.5.2 Model error and uncertainty

All experiments contain uncertainty, so there is uncertainty in the measurements you have taken. The experiment set-up (you may have assumed no load eccentricity, whereas, in reality, a small amount may occur), how the specimen interacts with the experiment (for example, the natural frequency of a support frame) may affect the results. As well as uncertainty around the tests themselves (for example, the natural variation in density of a material), we need to state and try to define the uncertainty in what we can/will control (the experiment set-up, the accuracy of the instruments, etc.) so that we can separate the real variation in the outcome from the variation due to our set-up. Hopefully, if the experiment is well-designed, this will be minimised, but we should always try to define and account for it. Model uncertainty is a huge subject, beyond the scope of this book; instead we recommend referring to Taylor's *An introduction to error analysis*[C.2].

Sometimes, our experiment may contain errors. An instrument may stop working, reach the end of its measurement range or become dislodged. The set-up may have been outside the intended tolerance. We were once perplexed by the range of data we had received for some rammed earth cubes (they were cubes rather than cylinders as the contractor had made them on-site and sent them for testing). It was only when we enquired that we discovered that some were tested parallel and some perpendicular to the earth compaction layers, leading to very different values. In these cases, it is important to determine what went wrong, what we can still know (for example if an instrument comes to the end of its range we may still have information on the elastic part of the response) and what we can no longer trust. As we discuss in Section C.5.3, if it is not clear what has gone wrong, we should be very cautious in assuming an erroneous result and should explore the possibility that a different result has been encountered.

As part of uncertainty, you will want to understand that your instrument is well-calibrated (Section C.2.3) and what the expected variation in instrument reading is.

C.5.3 Bias

For many experiments, as in the case of material strength or stress-strain tests, data collection and analysis is a largely objective process. However, if your tests are aimed at understanding the behaviour of a more complex system, a degree of understanding and interpretation is required. This is where bias can creep in.

We all have beliefs and opinions and, as we have emphasised previously, it's important to have an idea of what your test is trying to discover before you carry it out. These beliefs can sometimes be problematic as they may lead you to make decisions which are biased, and you may overlook relevant information. This is known as 'confirmation bias'.

It is human nature to infer an outcome from statistics that support their beliefs and understanding. Where fact-based decisions need to be made, confirmation bias can be a serious issue and you should take measures to avoid it.

This can occur in several ways, from the type and position of instruments we include in the experiment set-up or the type of experiment we choose to perform, to the way we interpret the outcome (and suggest different outcomes as outliers or bad data points). For example, if you were expecting a bi-linear response to the bamboo test in Figure C.4, you might assume D is the best result and be sceptical of the validity of some of the others. If you expected a parabolic response, F may appear to be the most reasonable response, and again, you may explain away other data sets as incorrectly set up or not showing the area of interest.

Explore techniques that allow you to challenge your observations. Obtain information from a range of sources. This could mean multiple repeats of the same test or a different test entirely that looks for similar outcomes. Consider and analyse the data from multiple perspectives and using different techniques.

Discuss your findings with a diverse group of people. This should include those who are experts in the field together with people with very little prior knowledge or understanding. Try to play devil's advocate, or you could even assign a team member to play this role.

'Anchoring' is a bias that leads to final judgements based on information gained early in the decision-making process. Anchoring can happen if you feel under pressure to make a quick decision (perhaps due to time or budget constraints) or if you are excited to reveal the outcome of your tests.

You should monitor your test during the process but try to avoid jumping to conclusions about the results until you have collected all the data and analysed it. Work through your plan methodically and in sequence. Make time to come to understand your findings and draw conclusions slowly. As you observe the tests, write down questions, rather than outcomes: "Just heard a loud bang, could this coincide with delamination?" You can then re-examine the questions and make sense of them alongside the rest of the data in due course.

With 'groupthink', the tendency is to form an opinion or take action because others have already expressed their position. This can be particularly difficult to avoid where those other opinions come from more senior or more experienced colleagues and collaborators. Try to gather your own thoughts, theories and questions, encourage others to do the same, and then compare. When you compare, start with the least senior/experienced team member and work up. It is harder to say something if someone more experienced has already said something that would contradict it. Try not to write-off the observation or question that one person had, but others didn't. Instead, hold the tension of multiple interpretations. Above all else, we are trying to design safely, so if we are unsure of something, rather than dismiss it, consider how it could be accounted for in the design.

Remember, there is no such thing as a stupid question. Ensure you involve a group of diverse individuals who feel empowered and confident to express opinions based on their own background and experience. You could go as far as using techniques like the Delphi method[C.3].

As a general rule, ask yourself and those involved in the tests, standardised, technically objective questions. What else could cause this behaviour? Is this failure mechanism unique to the test or would it be replicated in other situations? Ensure consistency in your approach. If a test is to be repeated, ensure all factors which can be controlled are controlled, to allow you to rule them out in the event of variations. Introduce objectivity into your decision-making and data analysis, and allow more time for it.

Always include **all** data from your testing unless there are good reasons to exclude any. It is particularly tempting to disregard data at the extremes of our expectations as 'freak' or 'erroneous'. Take time to understand the mechanism that led to those results and objectively question whether they can be ignored. Remember structural failure is generally an extreme event, so if this outcome occurred due to a failure in the test, could this failure be replicated on-site?

C.5.4 Design values

Once all the data has been gathered, analysed, questioned, reviewed and re-analysed, you will likely end up with a selection of data. These could be strengths, stiffness, shrinkage, aroma or colour. You will now want to try to use them in the design. We may already have an intended probability distribution or be trying to ascertain which would be most appropriate. We can use a statistical test to determine whether a normal or lognormal distribution is suitable. Depending on the distribution type, we can then calculate the appropriate values using prescribed methods. For example, for a normal distribution, we can calculate the mean and standard deviation. We can apply these values in our reliability analysis in accordance with Annex D of *Eurocode. Basis of structural design* or, if we are working on a known property, we can refer to Annex C[C.4].

If the values are critical for the safety of the building (or have a high risk for other reasons) it is recommended that once you have completed your own analysis you have the results verified by a third party. You will need to assess this on a project-by-project basis.

REFERENCES

C.1 National Physical Laboratory (NPL) website. Available at: https://www.npl.co.uk/ [Accessed: January 2025].

C.2 Taylor J.R. *An Introduction to Error Analysis: The Study of Uncertainties in Physical Measurements (Second edition).* Sausalito: University Science Books, 1997.

C.3 Salkey, N. and Helmer, O. *An Experimental Application of the Delphi Method to the Use of Experts.* Santa Monica: The Rand Corporation, 1962.

C.4 British Standards Institution. *BS EN 1990:2002+A1:2005: Eurocode. Basis of structural design. Annex C and D.* London: BSI, 2005.

APPENDIX D

Future imaginings

In writing this book we approached artist James McKay, who has worked previously in visualising future imaginings, to vividly represent this future. The images that he has created are rich in detail so here we simply list some of the future ideas that these images capture.

Construction site (present to near future):

- Diverse group of people, both labourers and designers/managers
- Everyone should be in protective equipment (hi-vis clothing, gloves, goggles, hard hat, steel-toe boots)
- Mix of ethnic clothing; you can, by law, wear a turban rather than hard hat, for example
- The operatives should have high tech tools — tablet/augmented reality, to ensure everything is placed precisely
- All items should have a QR code to ensure everything is placed in the right location
- An erected sign reporting:
  - Accidents — number of days since last accident/this site is accident-free etc.
  - Waste — tonnage of waste in the last week, and questioning how we can reduce it, aiming for zero waste

- Someone with a tablet checking the quality to ensure QA compliance for reduced factors of safety
- Sensors being installed to monitor current- and post-construction performance. These are attached to a data collection system — like a single bank from a data server
- A tower crane — with the construction company on it to include the word cooperative, suggesting a community construction company
- A concrete pour is a busy time — lots of people working at once
- Avoid too many machines doing the work!
- In the background a clear layout area with materials well organised, including reuse of elements
- Welfare facilities including gender neutral toilets

Engineering office:

- Diverse team around the table working on prototypes
- One is piloting a drone just above the table to take photos/laser scan the prototype
- Behind is a 'stress testing' rig
- Outside is a streetscape with green roofs, solar towers, large-scale solar shades/solar skins, green walls, green roofs and lots of birds (including parakeets)

Reuse yard:

- A variety of notices capturing changes in mindset around reuse
- Workers include Charity who is directing operations in the mid-ground
- A staff member using an exo-skeleton to pick up large weights
- A futuristic tractor/forklift
- Materials include lots of timber and stacks of stuff
- Timber sheds based on Yes Make style barns, bird boxes with swallows flying, green roofs, solar panels

Future street scene:

- Sustainable transport and cyclists, bike racks, electric or cyclo-taxis/vans
- Informal buildings (a kind of marketplace)
- Large-scale solar shades/solar skin umbrellas
- Lots of greenery, green roofs, green walls, wind turbines on the horizon, shutters on windows, informal cladding on the flats, solar panels on roofs, diverse community, wildflowers, wildlife (swifts, bees)